# Samsung Galaxy® S22

## by Bill Hughes

for **dummies**®

A Wiley Brand

## Samsung Galaxy® S22 For Dummies®

Published by: **John Wiley & Sons, Inc.,** 111 River Street, Hoboken, NJ 07030-5774, www.wiley.com

Copyright © 2022 by John Wiley & Sons, Inc., Hoboken, New Jersey

Published simultaneously in Canada

For general information on our other products and services, please contact our Customer Care Department within the U.S. at 877-762-2974, outside the U.S. at 317-572-3993, or fax 317-572-4002. For technical support, please visit https://hub.wiley.com/community/support/dummies.

Wiley publishes in a variety of print and electronic formats and by print-on-demand. Some material included with standard print versions of this book may not be included in e-books or in print-on-demand. If this book refers to media such as a CD or DVD that is not included in the version you purchased, you may download this material at http://booksupport.wiley.com. For more information about Wiley products, visit www.wiley.com.

Library of Congress Control Number: 2022935715

ISBN 978-1-119-87306-8 (pbk); ISBN 978-1-119-87307-5 (ebk); ISBN 978-1-119-87308-2 (ebk)

SKY10034163_042222

# Contents at a Glance

# Table of Contents

# Introduction

The Samsung Galaxy S22, S22+, and S22 Ultra are powerful smartphones, among the most powerful mobile phones ever sold. As of the publication of this book, the Galaxy S22s are the standard against which all other Android-based phones are measured.

Each cellular carrier offers a slightly customized version of the Galaxy S22 lineup. Some phones from cellular carriers come out of the box with preloaded applications, games, or files. This book doesn't dwell on these kinds of differences.

The name for each network is different, these phones are largely the same. (At least one marketing person at each cellular carrier is cringing as you read this.) This similarity allows me to write this book in a way that covers the common capabilities.

At a more core level, these phones are built for high-speed wireless communications, in particular the 5G networks you're seeing in ads. The cellular carriers have spent kajillions upgrading their networks to offer more coverage and better data speeds than their competition. Again, this book doesn't dwell on these differences in network technology because they don't really make much difference in ways that you can see in a book. (Again, at least one engineering person at each cellular carrier is cringing as you read this.)

Similarly, most of the capabilities among the different Galaxy S22 models are similar. The S22+ has a bigger screen and a bigger battery than the S22. Similarly, the S22 Ultra has a bigger screen with higher resolution, a bigger battery, and more camera capabilities than its little brothers. Actually, that's putting it mildly. The S22 Ultra has oh-my-goodness-you-cannot-be-serious kind of camera capabilities compared to every other mobile phone in the known universe. Otherwise, the three versions of the Galaxy S22 are practically identical. When there is an important distinction between the S22, the S22+, and the S22 Ultra, I mention it. Otherwise, I just call the phone the Galaxy S22 or just S22.

I assume that you already have a Galaxy S22, and I just hope that you have good coverage where you spend more of your time with your phone. If so, you'll be fine. If not, you need to switch to another network; otherwise, the experience with your phone will be frustrating. I would advise you to return your phone to that carrier and buy your Galaxy S22 at another cellular carrier. As long as you have good cellular data coverage, owning a Samsung Galaxy S22 will be an exciting experience!

First, in much the same way that different brands of PCs are all based on the Microsoft Windows operating system, all Galaxy S phones use the Google Android platform. The good news is that the Android platform has proven to be widely popular, even more successful than Google originally expected when it first announced Android in November 2007. More people are using Android-based phones, and more third parties are writing applications. This is good news because it offers you more options for applications (more on this in Chapter 8 on the Play Store, where you buy applications).

In addition, all Galaxy S22 phones use a powerful graphics processor, employ Samsung's Super AMOLED touchscreen, and are covered in Corning's Gorilla Glass. The superior screen differentiates this product line from other Android phones. Because of these enhanced capabilities, you can navigate around the screen with multi-touch screen gestures with ease. Plus, the videos look stunning from many angles.

Smartphones are getting smarter all the time, and the Galaxy S22 is one of the smartest. However, just because you've used a smartphone in the past doesn't mean you should expect to use your new Galaxy S22 without a bit of guidance.

You may not be familiar with using a multi-touch screen, and your new phone offers a lot of capabilities that you may or may not be familiar with. There used to be a physical button on the front to bring you back to the Home screen. It's no longer a physical button; instead, it's now software based. It would be unfortunate to find out from a kid in the neighborhood that the phone you've been carrying around for several months could solve a problem you've been having because you were never told that the solution was in your pocket the whole time.

In fact, Samsung is proud of the usability of its entire Galaxy lineup — and proud that the user's manual is really just a quick start guide. You can find lots of instructions on the web. However, you have to know what you don't know to get what you want unless you plan to view every tutorial.

That's where this book comes in. This book is a hands-on guide to getting the most out of your Galaxy S22.

# About This Book

This book is a reference — you don't have to read it from beginning to end to get all you need out of it. The information is clearly organized and easy to access. You don't need thick glasses to understand this book. This book helps you figure out what you want to do — and then tells you how to do it in plain English.

Within this book, you may note that some web addresses break across two lines of text. If you're reading this book in print and want to visit one of these web pages, simply key in the web address exactly as it's noted in the text, pretending as though the line break doesn't exist. If you're reading this as an e-book, you've got it easy — just click the web address to be taken directly to the web page.

# Foolish Assumptions

You know what they say about assuming, so I don't do much of it in this book. But I do make a few assumptions about you:

>> **You have a Galaxy S22 phone.** You may be thinking about buying a Galaxy S22 phone, but my money's on your already owning one. After all, getting your hands on the phone is the best part!

>> **You're not totally new to mobile phones.** You know that your Galaxy S22 phone is capable of doing more than the average phone, and you're eager to find out what your phone can do.

>> **You've used a computer.** You don't have to be a computer expert, but you at least know how to check your email and surf the web.

# Icons Used in This Book

Throughout this book, I used *icons* (little pictures in the margin) to draw your attention to various types of information. Here's a key to what those icons mean:

**TIP**

This whole book is like one big series of tips. When I share especially useful tips and tricks, I mark them with the Tip icon.

**REMEMBER**

This book is a reference, which means you don't have to commit it to memory — there is no test at the end. But once in a while, I do tell you things that are so important that I think you should remember them, and when I do, I mark them with the Remember icon.

**WARNING**

Whenever you may do something that could cause a major headache, I warn you with the, er, Warning icon.

**TECHNICAL STUFF**

These sections provide a little more information than is necessary to use your phone. The hope is that these sections convey extra knowledge to help you understand what is going on when things go wrong, or at least differently than you might have expected.

# Beyond the Book

In addition to what you're reading right now, this product also comes with a free access-anywhere Cheat Sheet. To get to this Cheat Sheet, simply go to www. dummies.com and type **Samsung Galaxy S22 For Dummies Cheat Sheet** in the Search box.

# Where to Go from Here

You don't have to read this book from cover to cover. You can skip around as you like. For example, if you need the basics on calling, texting, and emailing, turn to Part 2. To discover more about photos, games, and apps, go to Part 4.

Many readers are already somewhat familiar with smartphones and won't need the basic information found in Parts 1 and 2. A reasonably astute mobile phone user can figure out how to use the phone, text, and data capabilities. Parts 1, 2, and 3 are not for those readers. For them, I recommend skipping ahead to the chapters in Parts 4 through 6.

Former iPhone users, on the other hand, are a special case. (First, welcome to the world of Android!) The reality is that the iPhone and Galaxy S series have very similar capabilities, but these functions are just done in slightly different ways and emphasize different approaches to the similar problems. iPhone users, don't worry if you find that this book spends a fair amount of time explaining capabilities with which you're familiar. You can read through those chapters quickly, focus on the *how* instead of the description of *what,* and bypass potential frustration.

Current Samsung Galaxy S10, S20, and S21 users are also a special case. The Samsung Galaxy S22 is very similar to the earlier Galaxy S phones in many ways. Galaxy S22 operates mostly like these earlier models, but has improvements in usability, power consumption, and performance. Plus, the camera has even more capabilities (if you can believe it!). If you're comfortable with the earlier Galaxy models and now have a Galaxy S22, Chapters 15 and beyond will be of interest to you.

The majority of readers of this book are actually very astute and get the fact that this book covers the basics of using the Samsung Galaxy S22. A subset of readers complain in Internet reviews that a *For Dummies* book is too basic. If you do this, people will know that you did not read the title. Be sure to read the title and avoid public embarrassment.

# 1

# Getting Started with the Samsung Galaxy S22

Chapter **1**

# Exploring What You Can Do with Your Phone

hether you want just the basics from a mobile phone (make and take phone calls, customize your ringtone, take some pictures, maybe use a Bluetooth headset) or you want your phone to be always by your side (a tool for multiple uses throughout your day), you can make that happen. In this chapter, I outline all the things your Samsung Galaxy S22 can do — from the basics to what makes Galaxy S22 phones different from the rest.

## Discovering the Basics of Your Phone

All mobile phones on the market today include basic functions, and even some entry-level phones are a little more sophisticated. Of course, Samsung includes all basic functions on the Galaxy S22 model. In addition to making and taking calls (see Chapter 3) and sending and receiving texts (see Chapter 4), the Galaxy S22 sports the following basic features:

>> **50MP digital camera:** This resolution is for the S22 and S22+. The S22 Ultra has a mind-boggling 108MP. The mere 50MP is far more than is needed for posting good-quality images on the Internet. It is about right for making 24-x-36 posters that are photo quality. There is also a front-facing camera with 10MP that is useful for videoconference calls and selfies along with some combination of specialty lenses, depending upon the model. Your phone has some amazing capabilities (see Chapter 9 for more information on cameras and photographs and Chapter 10 on videos).

>> **Ringtones:** You can replace the standard ringtone with custom ringtones that you download to your phone. You also can specify different rings for different numbers.

>> **Bluetooth:** The Galaxy S22 phone supports stereo and standard Bluetooth devices. (See Chapter 3 for more on Bluetooth.)

>> **High-resolution screen:** The Galaxy S22 and the S22+ have screen resolutions of 2,340 x 1,080 pixels. The S22 Ultra offers the highest-resolution touchscreen on the market (3,088 x 1,440 pixels).

>> **Easy-to-use touchscreen:** The Galaxy S22 phone offers a very slick touchscreen that's sensitive enough to allow you to interact with the screen accurately, but not so sensitive that it's hard to manage. In addition, it has an optional setting that adjusts the sensitivity for special circumstances, like when you want to use one hand or add a screen protector!

# Taking Your Phone to the Next Level: The Smartphone Features

In addition to the basic capabilities of any entry-level phone, the Galaxy S22, which is based on the popular Android platform for mobile devices, has capabilities associated with other smartphones, such as the Apple iPhone:

>> **Internet access:** Access websites through a web browser on your phone.

>> **Wireless email:** Send and receive email from your phone.

>> **Multimedia:** Play music and videos on your phone.

- >> **Contact Manager:** The Galaxy S22 lets you take shortcuts that save you from having to enter someone's ten-digit number each time you want to call or text a friend. In fact, the Contact Manager can track all the numbers that an individual might have, store an email address and photo for the person, and synchronize with the program you use for managing contacts on both your personal and work PCs!

- >> **Digital camcorder:** The Galaxy S22 comes with a built-in digital camcorder that records live video at a resolution that you can set, including 8K UHD (8K ultra-high definition is the resolution that is available on the newest televisions). If you want to future-proof your videos, you can even select 8K for when you obtain a TV that operates at this resolution.

- >> **Mapping and directions:** The Galaxy S22 uses GPS (Global Positioning System) along with other complementary positioning systems to tell you where you are, find local services that you need, and give you directions to where you want to go.

- >> **Fitness information:** The Galaxy S22 automatically tracks important health information within the phone and with external sensors.

- >> **Business applications:** The Galaxy S22 can keep you productive while you're away from the office.

I go into each of these capabilities in greater detail in the following sections.

## Internet access

Until a few years ago, the only way to access the Internet when you were away from a desk was with a laptop. Smartphones are a great alternative to laptops because they're small, convenient, and ready to launch their web browsers right away. Even more important, when you have a smartphone, you can access the Internet wherever you are — whether Wi-Fi is available or not.

The drawback to smartphones, however, is that their screen size is smaller than that of even the most basic laptop. On the Galaxy S22 phone, you can use the standard version of a website if you want. You can pinch and stretch your way to get the information you want. (See Chapter 2 for more information on pinching and stretching. For more information on accessing the Internet from your Galaxy S22 phone, turn to Chapter 7.)

To make things a bit easier, most popular websites offer an easier-to-use app that you can download and install on your phone. This is discussed in detail in Chapter 8. Essentially, the website reformats the information from the site so that it's easier to read and navigate in the mobile environment. Figure 1-1 compares a regular website with the app version of that website.

Full Web Page

Mobile App Home Page

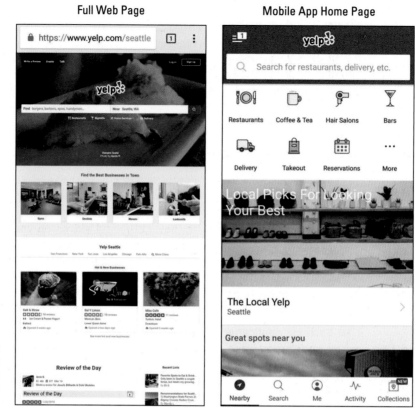

**FIGURE 1-1:**
A website
and the app
version of the
main site.

# Photos

The image application on your phone helps you use the digital camera on your Galaxy S22 phone to its full potential. It would almost make sense to call the Samsung Galaxy S22 a smart camera with a built-in phone! There are all kinds of smarts in these applications that automatically figure out what you're trying to do and make it so that you're suddenly the next Ansel Adams.

Studies have found that cellphone users tend to snap a bunch of pictures within the first month of phone usage. After that, the photos sit on the phone (instead of being downloaded to a computer), and the picture-taking rate drops dramatically.

The Galaxy S22 image management application is different. You can integrate your camera images into your home photo library, as well as photo-sharing sites such as Google Photos and Instagram, with minimal effort.

For more on how to use the Photo applications, turn to Chapter 9.

## Wireless email

On your Galaxy S22 smartphone, you can access your business and personal email accounts, reading and sending email messages on the go. Depending on your email system, you might be able to sync so that when you delete an email on your phone, the email is deleted on your computer at the same time so that you don't have to read the same messages on your phone and your computer.

Chapter 5 covers setting up your business and personal email accounts.

## Multimedia

Some smartphones allow you to play music and videos on your phone. On the Galaxy S22, you can use the applications that come with the phone, or you can download applications that offer these capabilities from the Play Store.

Chapter 12 covers how to use the multimedia services with your Galaxy S22 phone.

# Customizing Your Phone with Games and Applications

Application developers — large and small — are working on the Android platform to offer a variety of applications and games for the Galaxy S22 phone. Compared to the other smartphone platform, Google imposes fewer restrictions on application developers regarding what's allowable. This freedom to develop resonates with many developers — resulting in a bonanza of application development on this platform.

As of this writing, more than three million applications are available from Google's Play Store. For more information about downloading games and applications, turn to Chapter 8.

## Downloading games

Chapter 11 of this book is for gamers. Although your phone comes with a few general-interest games, you can find a whole wide world of games for every skill and taste. In Chapter 11, I give you all the information you need to set up different gaming experiences. Whether you prefer stand-alone games or multiplayer games, you can set up your Galaxy S22 phone to get what you need.

# Downloading applications

Your phone comes with some very nice applications, but these might not take you as far as you want to go. You might also have some special interests, such as philately or stargazing, that neither Samsung nor your carrier felt would be of sufficient general interest to include on the phone. (Can you imagine?)

Your phone also comes with preloaded *widgets*, which are smaller applications that serve particular purposes, such as retrieving particular stock quotes or telling you how your phone's battery is feeling today. Widgets reside on the extended Home screen and are instantly available.

Buying applications allows you to get additional capabilities quickly, easily, and inexpensively. Ultimately, these make your phone, which is already a reflection of who you are, even more personal as you add more capabilities.

# What's cool about the Android platform

The Samsung Galaxy S22 is a top-of-the-line Android phone. That means you can run any application developed for an Android phone to its full capability. This is significant because one of the founding principles behind the Android platform is to create an environment where application developers can be as creative as possible without an oppressive organization dictating what can and cannot be sold (as long as it's within the law, of course). This creative elbow room has inspired many of the best application developers to go with Android first.

## TAKE A DEEP BREATH

You don't have to rush to implement every feature of your Galaxy S22 phone the very first day you get it. Instead, pick one capability at a time. Digest it, enjoy it, and then tackle the next one.

I recommend starting with setting up your email and social accounts, but that's just me.

No matter how you tackle the process of setting up your Galaxy S22 phone, it'll take some time. If you try to cram it all in on the first day, you'll turn what should be fun into drudgery.

The good news is that you own the book that takes you through the process. You can do a chapter or two at a time.

In addition, Android is designed to run multiple applications at once. Other smartphone platforms have added this capability, but Android is designed to let you to jump quickly among the multiple apps that you're running — which makes your smartphone experience that much smoother.

# Surviving Unboxing Day

When you turn on your phone the first time, it will ask you a series of ten or so questions and preferences to configure it. Frankly, they are trying to make this book unnecessary and put me out of business. The nerve!

The good folks at Samsung are well-intentioned, but not every customer who owns a Samsung Galaxy S22 knows, from day one, whether they want a Samsung account, what's a good name for the phone, or what the purpose of a cloud service, such as Dropbox, is and how it would be used.

You can relax. I help you answer these questions — or, when appropriate, refer you to the chapter in this book that helps you come up with your answer.

On the other hand, if your phone is already set up, you probably took a guess or skipped some questions. Maybe now you're rethinking some of your choices. No problem. You can go back and change any answer you gave and get your phone to behave the way you want.

The following are the kinds of questions you may be asked. These questions may come in this order, but they may not. They typically include the following:

>> **Language/Accessibility:** This option lets you select your language. The default is English for phones sold within the United States. Also, the phone has some special capabilities for individuals with disabilities. If you have a disability and think you might benefit, take a look at these options. They have really tried to make this phone as usable as possible for as many folks as possible.

>> **Wi-Fi:** Your phone automatically starts scanning for a Wi-Fi connection. You can always use the cellular connection when you are in cellular coverage, but if there is a Wi-Fi connection available, your phone will try to use this first. It is probably cheaper and may be faster than the cellular.

At the same time, you may not want your phone to connect to the Wi-Fi access point with the best signal. It could be that the strongest signal is a fee-based service, whereas the next best signal is free. In any case, this page scans the available options and presents them to you.

>> **Date and Time:** This is easy. The default setting is to use the time and date that comes from the cellular network and the date and time format is the U.S. style. Just tap on the next button and move on. This date and time from the cellular network is the most accurate information you'll get, and you don't need to do anything other than be within cellular coverage now and again. If you prefer non-U.S. formatting, such as a 24-hour clock or day/month/year formatting, you can change your phone any way you want.

>> **Sign Up for a Samsung Account:** Go ahead and sign up for an account. The Samsung account offers you some nice things to help you get your phone back should you lose it. All you need is an account name, such as an email account, and a password.

**TIP**

When you buy a Galaxy S22 smartphone, you are now a customer of multiple companies! These include Samsung for the phone hardware, Google for the phone operating system (Android), and the wireless carrier that provides the cellular service. Plus, if you bought the phone through a phone retailer, such as Best Buy, it is in the mix as well. All of them want to make you happy, which is a good thing for the most part. The only downside is that they want to know who you are so that they can provide you with more services. Don't worry. You control how much they offer you.

>> **Google Account Sign-up:** *Google account* can means an email account where the address ends in @gmail.com. If you already have an account on Gmail, enter your user ID and password on your phone. If you don't have a Gmail account, I suggest waiting until you read Chapter 5. The good news is that you can use your existing email account. I highly recommend that you create a Google account, but it can wait until you read Chapter 5.

>> **Location Options:** Your phone knowing your location and providing it to an application can be sensitive issue.

If you're really worried about privacy and security, tap the green check marks on the screen and then tap the button that says Next. Selecting these options prevents applications from knowing where you are. (This choice also prevents you from getting directions and a large number of cool capabilities that are built into applications.) The only folks who'll know your location will be the 911 dispatchers if you dial them.

If you're worried about your security but want to take advantage of some of the cool capabilities built into your phone, tap the right arrow key to move forward. Remember, you can choose on a case-by-case basis whether to share your location. (I cover this issue in Chapter 17.)

>> **Phone Ownership:** This screen asks you to enter your first and last name. Go ahead and put in your real name. If you want to know more, read Chapter 5.

>> **Setting Up Voicemail:** Your cellular carrier manages the voicemail service that comes with your phone. If you're simply upgrading your phone but staying with the same carrier, you don't need to change your voicemail. If you're new to this carrier, you'll probably want to set up a new voicemail message. If you want to know more, read Chapter 3.

>> **Cloud Services:** The chances are that you will be offered the option to sign up for a cloud service where you can back up your phone and get access to a gazillion MB of free storage. This can be a tricky decision. You could sign up for every cloud service that comes along. Then you need to remember where you stored that critical file. You could sign up for one, and you may miss a nice capability that is available on another. You could have one cloud service for work and another for personal. Here is what I recommend: Sign up for whatever is the cloud service your phone offers during this initial setup process if you do not already have one. You will see what it is later. If you are happy with a cloud service you already have, such as Dropbox or OneDrive, chances are, they will have all the services you need for you and your phone. You can link your Galaxy S22 to this service by downloading the necessary app (which I cover how to do in Chapter 8).

>> **Learn about Key Features:** If you think you don't need this book, go ahead and take this tour of all the new things you can do. If you think you might need this book in any way, shape, or form, tap the Next button. This screen is for setting up the coolest and the most sophisticated capabilities of the phone. I cover many of them in this book.

>> **Device Name:** When this screen comes up, you'll see a text box that has the model name. You can keep this name for your phone, or you can choose to personalize it a bit. For example, you can change it to "Bill's Galaxy S22" or "Indy at 425-555-1234." The purpose of this name is for connecting to a local data network, as when you're pairing to a Bluetooth device. If this last sentence made no sense to you, don't worry about it. (I go over all of this in Chapter 3.) Tap Finish. In a moment, you see the Home screen, as shown in Figure 1-2.

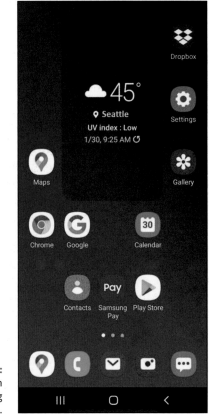

**FIGURE 1-2:**
The Home screen
for the Samsung
Galaxy S22.

Chapter **2**

# Beginning at the Beginning

I n this chapter, I fill you in on the basics of using your new Samsung Galaxy S22. You start by turning on your phone. (I told you I was covering the basics!) I guide you through charging your phone and getting the most out of your phone's battery. Stick with me for a basic tour of your phone's buttons and other features. Then I end by telling you how to turn off your phone or put it in "sleep" mode.

**TIP**

Unless you're new to mobile phones in general — and smartphones in particular — you might want to skip this chapter. If the term "smartphone" is foreign to you, you probably haven't used one before, and reading this chapter won't hurt. And, just so you know, a *smartphone* is just a mobile phone on which you can download and run applications that are better than what comes preloaded on a phone right out of the box.

# First Things First: Turning On Your Phone

When you open the box of your new phone, the packaging will present you with your phone, wrapped in plastic, readily accessible. If you haven't already, take the phone out of the box and remove any protective covering material on the screen. (There is no need to keep the plastic. It is not an effective screen protector!)

First things first. The Power button is on the right side of the phone. You can see where in Figure 2-1. Press the Power button for a second and see whether it vibrates and the screen lights up. Hopefully, your phone arrived with enough electrical charge that you won't have to plug it into an outlet right away. You can enjoy your new phone for the first day without having to charge it.

Power button

**FIGURE 2-1:**
The Power button on the Galaxy S22.

The phones that you get at the stores of most cellular carriers usually come with the battery partially charged and registered with the network.

## THE NITTY-GRITTY OF HOW YOUR PHONE WORKS

As soon as you turn on your phone, several things happen. As the phone is powering up, it begins transmitting information to (and receiving information from) nearby cellular towers. The first information exchanged includes your phone's electronic serial number. Every cellphone has its own unique serial number built into the hardware of the phone; the serial number in current-generation cellphones can't be duplicated or used by any other phone.

This electronic serial number is long. Cellular equipment is fine with long numbers. We mere mortals have enough trouble remembering 10-digit numbers, even with the hack of having only a limited number of area codes. To help us out, you and I get to use the shorter number and the cellular equipment happily keeps track of the 10-digit and the long device serial numbers and shows us only what we can handle.

It doesn't matter to the phone or the cellular tower if you're near your home when you turn on your phone — and that's the joy of mobile phones. All cellular networks have agreements that allow you to use cellular networks in other parts of the country and, sometimes, around the world.

That said, a call outside your cellular provider's own network may be expensive. Within the United States, many service plans allow you to pay the same rate if you use your phone anywhere in the United States to call anywhere in the United States. If you travel outside the United States, even to Canada, you might end up paying through the nose. *Remember:* Before you leave on a trip, check with your cellular carrier about your rates. Even if you travel internationally only a few times yearly, a different service plan may work better for you. Your cellular carrier can fill you in on your options.

**REMEMBER**

If the screen does light up, don't hold the Power button too long, or the phone might turn off.

If the phone screen doesn't light up (rats!), you need to charge the battery. That means that you have to wait to use your beautiful new phone. Sorry.

# Charging Your Phone and Managing Battery Life

To charge your phone, you need a battery charger. The most basic kind is a two-piece battery charger (cable and the transformer), as shown in Figure 2-2. (These used to come with the phone, but that's no longer the case.) Battery chargers are readily available at the store where you bought your phone.

**FIGURE 2-2:** The transformer and USB cable for charging your phone.

The cable has two ends: one end that plugs into the phone, and the other that's a standard rectangular USB connector. The phone end is a small connector called a *USB-C* that is used on many Samsung devices and is the standard for charging cellphones and other small electronics — and for connecting them to computers.

**TIP**

One cool thing you will notice about the USB-C connector is that you can insert it either side up. The older, rectangular USB-A connector probably drove you crazy because it wouldn't go in if you had the plug upside down. You still have that problem with the wall charger and cable, but you won't have the problem with the cable and phone!

To charge the phone, you have two choices:

>> Plug the transformer into a wall socket and then plug the cable's USB plug into the USB receptacle in the transformer. This is the option I recommend.

>> Plug the USB on the cable into a USB port on your PC.

Then you plug the small end of the cable into the phone. The port is on the bottom of the phone. Make sure that you push the little metal plug all the way in.

It doesn't really matter in what order you plug in things. However, if you use the USB port on a PC, the PC needs to be powered on for the phone to charge.

Your phone will charge faster with a Samsung wall charger that has Adaptive Fast Charging on it. When the wall charger or car charger has Adaptive Fast Charging, it knows if the phone is getting too warm and slows down the speed of charging. Because the charging speed can slow down, it can go faster the rest of the time.

If your phone is off when you're charging the battery, an image of a lightning bolt and the percentage charge level of the battery surrounded by a green circle appear onscreen for a moment. You can get the image to reappear with a quick press of the Power button. This image tells you the status of the battery without your having to turn on the phone.

If your phone is on, you see a small battery icon at the top of the screen showing how much charge is in the phone's battery. When the battery in the phone is fully charged, it vibrates to let you know that it's done charging.

Depending upon the circumstances, it can take less than an hour to go from a dead battery to a fully charged battery. You don't need to wait for the battery to be fully charged. You can partially recharge and run if you want.

You'll hear all kinds of "battery lore" left over from earlier battery technologies. For example, Lithium–polymer (LiPo) batteries in your S22 don't have a "memory" (a bad thing for a battery) as nickel–cadmium (NiCad) batteries did. That means that you don't have to make sure that the battery fully discharges before you recharge it. In fact, you want to avoid fully discharging the battery if you can avoid it.

In addition to the transformer and USB cable that come with the phone, you have other optional charging tools:

>> **USB travel charger:** If you already have a USB travel charger, you can leave the transformer at home. This accessory will run you about $15. You still need your cable, although any USB-to-USB-C cable should work.

>> **Car charger:** You can buy a charger with a USB port that plugs into the power socket/cigarette lighter in a car. This is convenient if you spend a lot of time in your car. The list price is $30, but you can get the real Samsung car charger for less at some online stores.

>> **Portable external charger:** You can buy a portable external charger that you can use to recharge your phone without having to plug into the power socket or cigarette lighter in a car. You charge this gizmo before your travel and connect it only when the charge in your phone starts to get low. These usually involve rechargeable batteries, but some of these products use photovoltaic cells to transform light into power. As long as there is a USB port (the female part of the USB), all you need is your cable. These chargers can cost from $30 to $100 on up. See some options in Figure 2-3.

>> **Wireless charger:** This option is slick. You simply put your phone on a charging mat or in a cradle, and the phone battery will start charging! Your Galaxy S22 uses the widely adopted Qi standard (pronounced *chee*). This option saves you from having to plug and unplug your phone.

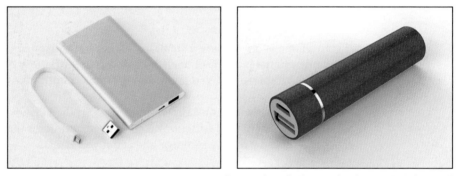

**FIGURE 2-3:** Some portable external charging options.

*Left: Stas Malyarevsky/Shutterstock, Right: GO DESIGN/Shutterstock*

**WARNING**

Ideally, use Samsung chargers or chargers from reputable manufacturers or retailers. The power specifications for USB ports are standardized. Reputable manufacturers comply with these standards, but less reputable manufacturers might not. Cheap USB chargers physically fit the USB end of the cable that goes to your phone. However, LiPo batteries are sensitive to voltage. There are many, many creative options available outside the store where you bought your phone, but avoid the allure of low price.

**WARNING**

Be aware that the conditions that make for a good charge with a photocell also tend to make for high heat. It will do you little good to have a beautifully function-ing charger and a dead phone.

It is best to get a charger with a USB-C Connector. All the cool kids have a device with a USB-C connector so there are many options in stores, but some of the chargers still out there are either for the Apple iPhone and have an Apple Light-ning connector or have a micro-USB connector. A micro USB will not work with a USB-C connector. You either need to get a cable that has a USB-C connector or use an adapter that will allow you to connect micro USB to a USB-C connector. Figure 2-4 shows the micro USB to USB-C adapter.

Yoyochow23/Shutterstock

**FIGURE 2-4:**
The micro
USB to USB-C
adapter.

## ONLY YOU CAN PREVENT PHONE FIRES

The Galaxy S22 uses lithium-polymer (LiPo) batteries. This is related to, but different from, Lithium-ion batteries. As the name implies, these batteries are based upon the element lithium (Li). Lithium works really well for cellphone batteries because it's very light.

The problem is that if lithium gets too hot, it burns. Lithium can catch fire at a relatively modest 180°F, and then burn like crazy at 1,100°F. By comparison, dry firewood won't ignite until 451°F, and will then burn at 1,100°F. In other words, lithium burns just as hot as a campfire, but it ignites at a much lower temperature.

To make things safe, battery makers and cellphone manufacturers implement all kinds of sneaky and imaginative methods of removing the heat from around the lithium battery. In addition, your phone invisibly monitors temperatures and takes steps to ensure that it doesn't get too hot. This all works so well and seamlessly that you probably didn't even notice this was happening in your earlier phones and other gizmos that use Lithium technology.

*(continued)*

*(continued)*

So, here's the big problem: If things start going wrong with Lithium-based batteries, many of the sneaky and imaginative methods used to remove the heat begin to fail. The fancy term for what happens next is *thermal runaway.* When thermal runaway happens in a device with a Lithium battery, one of two things happens:

- The guts of the phone will bubble and melt.

- The battery will burst into flames, throwing sparks for several feet in all directions.

LiPo is supposed to be safer than Li-ion, but a bunch of YouTube videos can show you what can happen when you create the wrong situation. So, what should you do and, perhaps more important, what should you *not* do? If your phone has started to melt, the best possible solution is to dunk it in a glass, bucket, or toilet full of water. If you can't find something to dunk it in, you can throw water on the phone to stop the fire. Water both cools the lithium and prevents oxygen from getting to the flame. I know: Getting your phone wet goes against all your instincts, but once a phone has started bubbling, all your texts and photos have already been lost. Prevent further damage and use water to put out the fire.

If you don't have water handy, and you're somewhere outside, in a parking lot or on a sidewalk away from people, you can set the phone on the ground, get far away from it, and call 911. Stand far back from the phone. It could explode. For obvious reasons, putting out the fire by dunking the phone in water or dousing it with water is best.

Smothering the phone with a blanket or dirt won't work. Even if you get the fire stopped momentarily with these methods, the battery is likely to start burning again because it's still too hot. A fire extinguisher with anything other than water or a liquid will also not work well. As with a blanket or dirt, the battery stays hot, and when oxygen reaches it, the lithium will simply reignite.

## SHARING IS CARING: WIRELESS PowerShare

A great deal of attention is paid to keeping your phone charged. Sometimes your phone has plenty of power, but another device is running low. Your S22 has a cool new feature called Wireless PowerShare. You go into Settings, turn this capability on, and the back of your S22 becomes a wireless charger! You can charge the Galaxy S20 of your best friend and be a hero for the day.

# Navigating the Galaxy S22

Galaxy S22 phone devices differ from earlier mobile phones in design: They have significantly fewer hardware buttons (physical buttons on the phone). They rely much more heavily on software buttons that appear onscreen.

In this section, I guide you through your phone's buttons.

## The phone's hardware buttons

Samsung has reduced the number of hardware buttons on the Galaxy S22. There are only three: the Power button, the Volume Down button, and the Volume Up button. Before you get too far into using your phone, orient yourself to be sure that you're looking at the correct side of the phone. When I refer to the *left* or *right* of the phone, I'm assuming a vertical orientation (meaning you're not holding the phone sideways) and that you're looking at the phone's screen.

### The Power button

The Power button (refer to Figure 2-1) is on right side of the phone, toward the middle when you hold it in vertical orientation.

In addition to powering up the phone, pressing the Power button puts the device into sleep mode if you press it for a moment while the phone is On. *Sleep mode* shuts off the screen and suspends most running applications.

The phone automatically goes into sleep mode after about 30 seconds of inactivity to save power, but you might want to do this manually when you put away your phone. The Super AMOLED (Active-Matrix Organic Light-Emitting Diode) screen on your Samsung Galaxy S22 is cool, but it also uses a lot of power.

**REMEMBER**

Don't confuse sleep mode with powering off. Because the screen is the biggest user of power on your phone, having the screen go blank saves battery life. The phone is still alert to any incoming calls; when someone calls, the screen automatically lights up.

### The Volume button(s)

Technically, there are two Volume buttons: one to increase the volume, and the other to lower it. Their locations are shown in Figure 2-5.

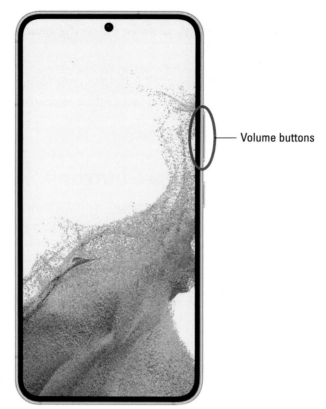

Volume buttons

**FIGURE 2-5:**
The Galaxy
S22 Volume
buttons on the
upper right.

The Volume buttons control the volume of all the audio sources on the phone, including:

>> The phone ringer for when a call comes in (ringtone)

>> The notifications that occur only when you're not talking on the phone, such as the optional ping that lets you know you've received a text or email

>> The phone headset when you're talking on the phone

>> The volume from the digital music and video player (media)

The volume controls are aware of the context; they can tell which volume you're changing. For example, if you're listening to music, adjusting volume raises or lowers the music volume, but leaves the ringer and phone-earpiece volumes unchanged.

The Volume buttons are complementary to software settings you can make within the applications. For example, you can open the music-player software and turn up the volume on the appropriate screen. Then you can use the hardware buttons to turn down the volume, and you'll see the volume setting on the screen go down.

Another option is to go to a settings screen and set the volume levels for each scenario. Here's how to do that:

1. **From the Home screen, press either Volume button.**

   You can press it either up or down. Doing so brings up the screen shown in Figure 2-6.

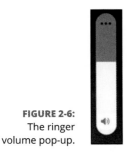

**FIGURE 2-6:**
The ringer
volume pop-up.

If you press the Volume Up or Volume Down button, the ring tone gets louder or softer. Hold off on this tweak for now and go to the next step.

2. **From this screen, tap the three dots at the top.**

   Tapping it brings up the screen shown in Figure 2-7.

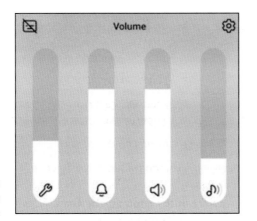

**FIGURE 2-7:**
The All Volume
Settings pop-up.

3. **From the screen shown in Figure 2-7, set the volume at the desired setting.**

   You can adjust the volume of any setting by placing your finger on the slider. You can slide it down to lower this particular volume setting or up to raise it.

# The touchscreen

To cram all the information that you need onto one screen, Samsung takes the modern approach to screen layout. You'll want to become familiar with several finger-navigation motions used to work with your screen.

Before diving in, though, here's a small list of terms you need to know:

>> **Icon:** An *icon* is a little image. Tapping an icon launches an application or performs some function, such as making a telephone call.

>> **Button:** A *button* on a touchscreen is meant to look like a three-dimensional button that you would push on, say, a telephone. Buttons are typically labeled to tell you what they do when you tap them. For example, you'll see buttons labeled Save or Send.

>> **Hyperlink:** Sometimes called a *link* for short, a *hyperlink* is text that performs some function when you tap it. Usually text is lifeless. If you tap a word and it does nothing, then it's just text. If you tap a word and it launches a website or causes a screen to pop up, it's a hyperlink.

>> **Thumbnail:** A *thumbnail* is a small, low-resolution version of a larger, high-resolution picture stored somewhere else.

With this background, it's time to discuss the motions you'll be using on the touchscreen.

**WARNING**

You need to clean the touchscreen glass from time to time. The glass on your phone is Gorilla Glass (made by Corning) — the toughest stuff available to protect against breakage. Use a soft cloth or microfiber to get fingerprints off. You can even wipe the touchscreen on your clothes. However, never use a paper towel! Over time, glass is no match for the fibers in the humble paper towel.

## Tap

Often, you just tap the screen to make things happen (as when you launch an app) or select options. Think of a *tap* as a single mouse click on a computer screen. A tap is simply a touch on the screen, much like using a touchscreen at a retail kiosk. Figure 2-8 shows what the tap motion should look like.

One difference between a mouse click on a computer and a tap on a Galaxy S22 phone is that a single tap launches applications on the phone in the same way that a double-click of the mouse launches an application on a computer.

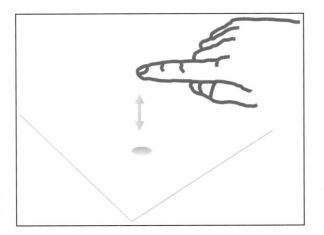

**FIGURE 2-8:**
The tap
motion.

TIP

A tap is different from press and hold (see the next section). If you leave your finger on the screen for more than an instant, the phone thinks you want to do something other than launch an application.

## Press and hold

*Press and hold,* as the name implies, involves putting your finger on an icon on the screen and leaving it there for more than a second. What happens when you leave your finger on an icon depends upon the situation.

For example, when you press and hold on an application on the Home screen (the screen that comes up after you turn on the phone), a garbage-can icon appears onscreen. This is to remove that icon from that screen. And when you press and hold an application icon from the list of applications, the phone assumes that you want to copy that application to your Home screen. Don't worry if these distinctions might not make sense yet. The point is that you should be familiar with holding and pressing — and that it's different from tapping.

TIP

You don't need to tap or press and hold very hard for the phone to know that you want it to do something. Neither do you need to worry about breaking the glass, even by pressing on it very hard. If you hold the phone in one hand and tap with the other, you'll be fine. I suppose you might break the glass on the phone if you put it on the floor and press up into a one-fingered handstand. I don't recommend that, but if you do try it, please post the video on YouTube.

WARNING

On average, a person calls 911 about once every year. Usually, you call 911 because of a stressful situation. Like every phone, the Samsung Galaxy S22 has a special stress sensor that causes it to lock up when you need it most. Okay, not really, but it seems that way. When you're stressed, it's easy to think that you're tapping when you're actually pressing and holding. Be aware of this tendency and remember to *tap.*

## Moving around the screen or to the next screen

Additional finger motions help you move around the screens and to adjust the scaling for images that you want on the screen. Mastering these motions is important to getting the most from your phone.

The first step is navigating the screen to access what's not visible onscreen. Think of this as navigating a regular computer screen, where you use a horizontal scroll bar to access information to the right or left of what's visible on your monitor, or a vertical scroll bar to move you up and down on a screen.

The same concept works on your phone. To overcome the practical realities of screen size on a phone that will fit into your pocket, the Galaxy S22 phone uses a *panorama* screen layout, meaning that you keep scrolling left or right (or maybe up and down) to access different screens.

In a nutshell, although the full width of a screen is accessible, only the part bounded by the physical screen of the Galaxy S22 phone is visible on the display. Depending upon the circumstances, you have several ways to get to information not visible on the active screen. These actions include drag, flicks, pinch and stretch, and double taps. I cover all these gestures in the following sections.

## Drag

The simplest finger motion on the phone is the *drag.* You place your finger on a point on the screen and then drag the image with your finger. Then you lift your finger. Figure 2-9 shows what the motion looks like.

**FIGURE 2-9:** The drag motion for controlled movement.

Dragging allows you to move slowly around the panorama. This motion is like clicking a scroll bar and moving it slowly.

## Flick

To move quickly around the panorama, you can *flick* the screen to move in the direction of your flick (see Figure 2-10).

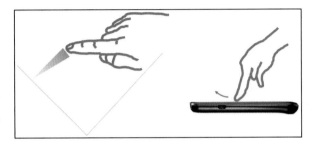

Better control of this motion comes with practice. In general, the faster the flick, the more the panorama moves. However, some screens (such as the extended Home screen) move only one screen to the right or left, no matter how fast you flick.

## Pinch and stretch

Some screens allow you to change the scale of images you view on your screen. When this feature is active, the Zoom options change the magnification of the area on the screen. You can zoom out to see more features at a smaller size or zoom in to see more detail at a larger size.

To zoom out, you put two fingers (apart) on the screen and pull them together to pinch the image. Make sure you're centered on the spot that you want to see in more detail. The pinch motion is shown in Figure 2-11.

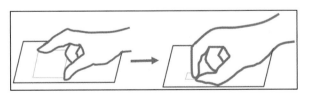

**FIGURE 2-11:**
Use the pinch
motion to
zoom out.

The opposite motion is to zoom in. This involves the stretch motion, as shown in Figure 2-12. You place two fingers (close together) and stretch them apart.

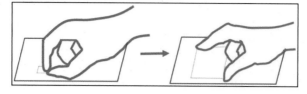

**FIGURE 2-12:**
Use the stretch
motion to
zoom in.

## Double tap

The *double tap* (shown in Figure 2-13) just means tapping the same button area on the screen twice in rapid succession. You use the double tap to jump between a zoomed-in and a zoomed-out image to get you back to the previous resolution. This option saves you frustration in getting back to a familiar perspective.

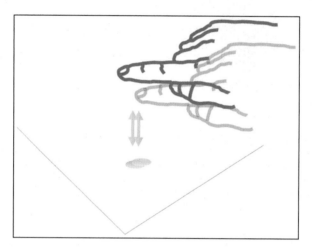

**FIGURE 2-13:**
The double-tap
motion.

**REMEMBER**

When you double tap, time the taps so that the phone doesn't interpret them as two separate taps. With a little practice, you'll master the timing of the second tap.

## The extended Home screen

The Home screen is the first screen you see when the phone is done with setting up. Additional screens off to the right and left make up the extended Home screen. They can be seen as a panorama in Figure 2-14.

At any given moment, you see only one screen at a time. You navigate among the screen by flicking to the right and left. The Home button, which I describe in more detail in the next section, will always bring you back to the Home screen.

Home Screen

**FIGURE 2-14:**
The Galaxy
S22 phone
panorama
display of
the extended
Home screen.

The extended Home screen is where you can organize icons and other functions to best make the phone convenient for you. Out of the box, Samsung and your cellular carrier have worked together to create a starting point for you. Beyond that, though, you have lots of ways you can customize your Home screens for easy access to the things that are most important to you. Much of the book covers all the things that the phone can do, but a recurring theme is how to put your favorite capabilities on your Home screen, if you want.

To start, check out the layout of the Home screen and how it relates to other areas of the phone. Knowing these areas is important for basic navigation.

Figure 2-15 shows a typical Home screen and highlights three important areas on the phone:

>> **The notification area:** This part of the screen presents you with small icons that let you know if something important is up, like battery life or a coupon from McDonald's.

>> **The primary shortcuts:** These four icons remain stationary as you move across the Home screen. If you notice in Figure 2-15, these have been determined by Samsung and your cellular carrier to be the four most important applications on your phone and are on all the screens.

>> **The Function keys:** These three keys control essential phone functions, regardless of what else is going on at the moment with the phone.

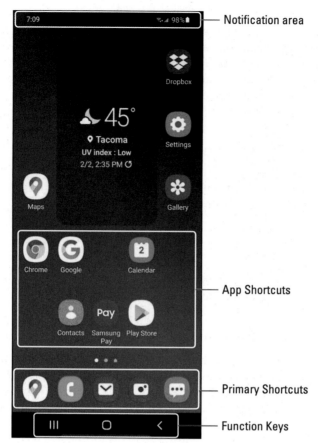

Notification area

App Shortcuts

Primary Shortcuts

Function Keys

**FIGURE 2-15:**
Important areas on the Galaxy S22 phone and Home screen.

**TIP**

There are a series of dots just above the primary shortcuts on the extended Home screen. You may also notice that one of the dots isn't just a dot — it's a little house. (It's very tiny, so you may not notice it at first, but in Figure 2-15, the house is to the right of the two dots.) That is the "home" Home screen. The brightest dot indicates where you are among the screens. You can navigate among screens by dragging the screen to the left or right. This moves you one screen at a time. You can also jump multiple screens by tapping on the dot that corresponds to the screen number you want to see or by dragging the dots to the screen you want to see. The following sections give you more detail on each area.

## Using the Home button

One of the important design features Samsung has implemented with the Galaxy phone is the "invisible" Home button. Some other smartphones have a physical button on the bottom of the front screen. To offer more screen real estate, the S22 makes it a software-based button on the bottom in the center of the function keys.

The Home button (see Figure 2-16) may not look like much, but it is very important because it brings you back to the Home screen from wherever you are in an application. If you're working on applications and feel like you're helplessly lost, don't worry. Press the Home button, close your eyes, click your heels together three times, and think to yourself, "There's no place like home," and you will be brought back to the Home screen.

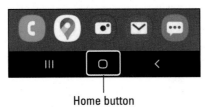

**FIGURE 2-16:**
The Galaxy S22
Home button
on the front.

Home button

**TIP**

You don't really need to do all that other stuff after pressing the Home button. Just pressing the Home button does the trick.

## Adding shortcuts to the Home screen

You have a lot of screen real estate where you can put icons of your favorite applications and widgets (refer to Figure 2-14). (*Widgets* are small apps that take care of simple functions, like displaying time or the status of your battery.) You can add shortcuts to the apps and to these widgets to your Home screen by following these steps:

**1.** **From the extended Home screen page where there is some space for the icon of an app, swipe your finger upward on the screen.**

This brings up a directory of all the apps you currently have on the phone. The number of apps and pages for their icons is practically unlimited. (I cover how to add new apps in Chapter 8.)

**2.** **Press and hold the icon of the app you want to add.**

The icon under your finger will get a pop-up like the one shown in Figure 2-17.

**FIGURE 2-17:**
An Apps page.

3. **Tap the Add to Home option.**

   The icon now appears in a random spot on one of your home pages. If you like where the icon is, you're all set. If you want to move it to another spot on your Home page, continue on to the next step.

4. **Press and hold the icon you want to move.**

   Two things happen with the icon. First you get a pop-up that asks you if you want to uninstall the app and some other options. To move the icon, slide your finger on the screen to where you want it to live. When you're happy with its location, lift your finger and it now has a new home.

Done.

## Taking away shortcuts

Say that you find that you really don't use that app as much as you thought you would, and you want to have less clutter on your Home page. No problem. Press and hold the icon. A pop-up like the one shown in Figure 2-18 appears.

FIGURE 2-18:
The shortcut pop-up on the Home page.

Simply tap the garbage can, and off it goes to its maker.

It's gone from your Home page, but if you made a mistake, you can get it back easily enough. To re-create it, simply go back to the App Menu key and follow the process again. It's still on your phone.

# The notification area and screen

The notification area is located at the top of the phone (refer to Figure 2-15). Here, you see little status icons. Maybe you received a text or an email, or you'll see an application needs some tending to.

Think of the notification area as a special email inbox where your carrier (or even the phone itself) can give you important information about what's happening with your phone. These little icons at the top tell you the condition of the different radio systems on your phone: The number of bars shown gives you an indication of signal strength, and usually the phone will also tell you what kind of mobile data speed you're getting — such as 3G, 4G, or 5G.

You could take the time to learn the meanings of all the little icons that might come up, but that would take you a while. A more convenient option is to touch the notification area and pull it down, as shown in Figure 2-19.

**FIGURE 2-19:**
Notification
area initial
pull-down.

This brings up the most commonly used notifications. In this case, it's the cellular signal strength indicator, sound, Bluetooth, auto-rotate, flashlight, and airplane mode. This information is written so that you can understand what's going on and what, if anything, you're expected to do — for example, if you see that you have connected to the Wi-Fi access point you intended.

If this screen gets too full, you can clear it by tapping the Clear button. You can also clear notifications one at a time by touching each one and swiping it to the side. If you want a shortcut to control some key phone capabilities, you can tap one of these icons. For example, if you want to turn off Wi-Fi, you can just tap the Wi-Fi icon. This saves you from having to go to Settings to do the same thing.

There are more handy capabilities at your disposal. When you're finished reading these notifications, slide your finger from top to bottom.

This brings up many more options that you would otherwise need to find with the Settings icon (see Figure 2-20). If the icon is blue, it means that the capability is enabled. If the icon is gray, it is not enabled.

FIGURE 2-20:
Notification area second pull-down.

# The Device Function keys

At the bottom of the screen are three important buttons: the Device Function keys. They're always ready for you to navigate your phone even though they might switch off to hide their presence and give you more screen. Whatever else you're doing on the phone, these buttons can take over. I talk about the Home button earlier in this chapter. The other buttons are equally cool.

## Recent Apps button

The button to the left of the Home button, the three vertical lines, is the Recent Apps button. Tapping the button brings up a list of the open apps. This is handy to navigate between running apps. Figure 2-21 shows a typical screen of recent apps.

You can scroll among the open apps by flicking left or right. You can jump to one of these apps by tapping somewhere on the image. You can also shut down one of the apps by flicking upward. You can shut them all down by tapping the link at the bottom that says Close All.

**FIGURE 2-21:**
Recent Apps
screen.

The Device Function keys are cool because they light up when you're touching them or the screen, but fade away the rest of the time.

**TIP**

### The Back button

The Back button on your phone is similar to the Back button in a web browser: It takes you back one screen.

As you start navigating through the screens on your phone, tapping the Back button takes you back to the previous screen. If you keep tapping the Back button, you'll eventually return to the Home screen.

## The keyboard

The screen of the Galaxy S22 phone is important, but you'll still probably spend more time on the keyboard entering data on the QWERTY keyboard.

## Using the software keyboard

The software keyboard automatically pops up when the application detects a need for user text input. The keyboard, shown in Figure 2-22, appears at the bottom of the screen.

## Using voice recognition

The third option for a keyboard is . . . no keyboard at all! Galaxy S22 phones come with voice recognition as an option. It's very easy and works surprisingly well. In most spots where you have an option to enter text, you see a small version of the microphone icon shown in Figure 2-23.

**FIGURE 2-23:**
The voice
recognition
icon.

Just tap this icon and say what you would have typed. You see the phone thinking for a second, and then a screen tells you that it is listening. Go ahead and start talking. When you're done, you can tap the Done button or just be quiet and wait. Within a few seconds, you'll read what you just said!

## The orientation of the phone

Earlier in this chapter where I discuss the Power button, I refer to the phone being in vertical orientation (so that the phone is tall and narrow). It can also be used in the *landscape* orientation (sideways, or so that the phone is short and wide). The phone senses in which direction you're holding it and orients the screen to make it easier for you to view.

REMEMBER

The phone makes its orientation known to the application, but not all applications are designed to change their inherent display. That nuance is left to the writers of the application. For example, your phone can play videos. However, the video player application that comes with your phone shows video in landscape mode only.

In addition, the phone can sense when you're holding it to your ear. When it senses that it's held in this position, it shuts off the screen. You need not be concerned that you'll accidentally "chin-dial" a number in Botswana.

# Going to Sleep Mode/Turning Off the Phone

You can leave your phone on every minute until you're ready to upgrade to the newest Galaxy phone in a few years, but that would use up your battery in no time. Instead, put your idle phone in sleep mode to save battery power. *Note:* Your phone goes into sleep mode automatically after 30 seconds of inactivity on the screen.

TIP

You can manually put the phone in sleep mode by pressing the Power button for just a moment.

Sometimes it's best to simply shut down the phone if you aren't going to use it for several days or more. To shut down the phone completely, pull down the initial notifications screen, swipe down again for the second notifications screen (refer to Figure 2-20), and press the Power Off button. The following options appear (see Figure 2-24):

>> **Power Off:** Shut down the phone completely.

>> **Restart:** The Android operating system is stable. Well, mostly stable. Sometimes you may want to reboot it because it starts acting quirky. Tap the restart option and see whether that solves the problem.

>> **Emergency Mode:** After you have agreed to some terms and conditions, your phone will go into semi-hibernation and take actions that limit battery use very severely. It can come back to life quickly if you feel you need to dial 911. Before you can use this option, you need to agree to some of the limitations. You can disable the mode by returning to the Power Off icon on the second notifications screen.

Good night!

Power off

Restart

**FIGURE 2-24:**
Options when
you tap the
Power Off
button.

Emergency mode
Off

# 2
# Communicating with Other People

# Chapter **3**

# Calling People

A t its essence, any cellphone — no matter how fancy or smart — exists to make phone calls. The good news is that making and receiving phone calls on your Galaxy S22 is easy.

In this chapter, I show you not only how to make a call, but also how to use your call list to keep track of your calls. And don't skip the section on using your phone for emergencies.

Finally, if you're like many people, you're never doing just one thing at a time, and a Bluetooth headset can make it easier for you to talk on the phone while driving and getting directions, checking email, wrangling kids and dogs, or just plain living life. In this chapter, I show you how to hook up your phone to a Bluetooth headset so that you can make and receive phone calls hands-free.

## Making Calls

After your phone is on and you're connected to your cellular carrier (see Chapters 1 and 2), you can make a phone call. It all starts from the Home screen. Along the bottom of the screen, above the Device Function keys, are either four or five icons, which are the *primary shortcuts* (see Figure 3-1). The primary shortcuts on your phone may differ slightly, but in this case, from left to right, they are

>> Phone

>> Camera

>> Email

>> Messages

**FIGURE 3-1:**
The primary shortcuts on the Home screen.

To make a call, follow these steps:

**1.** **From the Home screen, tap the Phone icon.**

You see a screen like the one shown on the left in Figure 3-2.

**2.** **Tap the telephone number you want to call.**

The Keypad screen (on the right in Figure 3-2) shows the numbers you've entered. As you type, it tries to guess who you're calling.

TIP

For long distance calls while in the United States, you don't need to dial 1 before the area code — just dial the area code and then the seven-digit phone number. Similarly, you can include the 1 and the area code for local calls. On the other hand, if you're traveling internationally, you need to include the 1 — and be prepared for international roaming charges!

In Chapter 6, you can read about how to make a phone call through your contacts.

**3.** **Tap the blue phone button at the bottom of the screen to place the call.**

The screen changes to the screen shown in Figure 3-3. You have a chance to verify you dialed the person you intended.

Within a few seconds, you should hear the phone ringing at the other end or a busy signal. From then on, it is like a regular phone call.

**4.** **When you're done with your call, tap the red phone button at the bottom of the screen.**

The call is disconnected.

Without Telephone Number

With Telephone Number

**FIGURE 3-2:**
Dial the
number from
the Keypad
screen.

If the other party answers the phone, you have a few options available to you by tapping on the correct icon/hyperlink on the screen, including:

» Put the call on hold.

» Add another call to have a three-way conversation.

» Increase the volume.

» Switch on a Bluetooth device (see the section "Syncing a Bluetooth Headset," later in this chapter).

» Turn on the phone's speaker.

» Bring up the keypad to enter numbers.

» Mute the microphone on the phone.

You can do any or all of these, or you can just enjoy talking on the phone.

If the call doesn't go through, either the cellular coverage where you are is insufficient, or your phone got switched to Airplane mode, which shuts off all the radios in the phone. (It's possible, of course, that your cellular carrier let you out the door without having set you up for service, but that's pretty unlikely!)

**FIGURE 3-3:**
Dialing screen.

Check the notification section at the top of the screen. If there are no connection-strength bars, try moving to another location. If you see a small airplane silhouette, bring down the notification screen (see how in Chapter 2) and tap the plane icon to turn off Airplane mode.

TIP

If you pull down the notification screen and don't see the silhouette of an airplane, scroll the green or gray icons downward a second time. The icon may be off the page.

# Answering Calls

Receiving a call is even easier than making a call. When someone calls you, caller ID information appears in a pop-up screen. Figure 3-4 shows some screen options for an incoming call.

Full Screen Incoming call                Partial Screen Incoming call

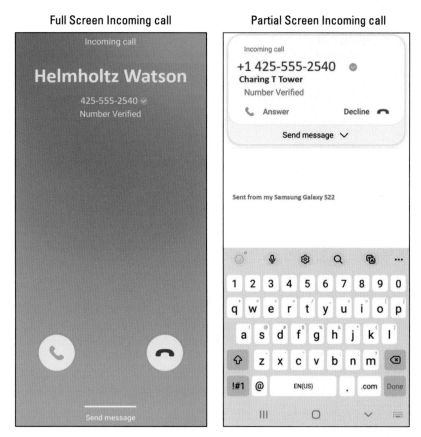

FIGURE 3-4:
Possible
screens
when you're
receiving a call.

If you aren't doing anything with the phone at that moment, it will present you with the full screen. If you're using an app, it will give you a pop-up screen, as shown in the image on the right. To answer the call, press and slide the blue phone button. To not answer a call, you can simply ignore the ringing, or you can press and slide the red phone button. The ringing stops immediately, and in either case, the call goes to voicemail.

**TIP**

In Part 4, I fill you in on some exciting options that you can enable (or not) when you get a call. For example, you can specify a unique ringtone for a particular number, or have an image of the caller pop up (if you save your contacts to your phone).

Regardless of what you were doing on the phone at that moment — such as listening to music or playing a game — the answer pop-up screen can appear. Any active application, including music or video, is suspended until the call is over.

For callers to leave you messages, you must set up your voicemail. If you haven't yet set up your voicemail, the caller will hear a recorded message saying that your voicemail account isn't yet set up. Some cellular carriers can set up voicemail for you when you activate the account and get the phone; others require you to set up voicemail on your own. Ask how voicemail works at your carrier store or look for instructions in the manual included with your phone.

Most cellular carriers give you a special seven-digit telephone number for your voicemail that should be in your contacts with the name Voicemail (Chapter 6 explains more about contacts). For now, the thing to remember is that you can call this number to record the message if you're unable to answer the call and to retrieve messages.

Don't give your voicemail number to anyone. If you do share it so that person can just call directly to your voicemail, you'll want to increase security and require a four-digit PIN before anyone can retrieve your messages.

The answer and reject icons are pretty standard on any cellular phone. However, your Galaxy S22 is no standard phone. There is a third option, and what happens depends on your individual phone. In addition to the standard options of answer or reject, you have one more option — to reject *and* send the caller a text message. As your caller is sent to your voicemail, you also can immediately send the caller a text message that acknowledges the call.

Some of the typical canned messages that you can send are:

>> Sorry, I'm busy. Call back later.

>> I'm in a meeting.

>> I'll call you back.

>> I'm at the movie theater.

>> I'm in class.

You tap the message that applies. The message is sent as a text right away, which alerts the caller that you're not ignoring them — it's just that you can't talk right now. Nice touch.

You can also create and store your own message, like "Go away and leave me alone," or "Whatever I am doing is more important than talking to you." You could also be polite. To create your own canned message, tap Compose new message and type away. It's then stored on your phone for when you need it.

**REMEMBER**

The caller has to be able to receive text messages on the phone used to make the call. This feature doesn't work if your caller is calling from a landline or a cellphone that can't receive texts.

# Keeping Track of Your Calls: The Recents

One of the nice features of cellular phones is that the phone keeps a record of the calls that you've made and received. Sure, you might have caller ID on your landline at home or work, but most landline phones don't keep track of whom you called. Cellphones, on the other hand, keep track of all the numbers you called. This information can be quite convenient, like when you want to return a call and you don't have that number handy. In addition, you can easily add a number to the contact list on your phone.

When you tap the Recents link on the phone screen, you get a list of all incoming and outgoing calls. (This hyperlink is located at the bottom of the screen, shown on the left side in Figure 3-2.) Tapping the link shows the list of recent calls, as shown in Figure 3-5.

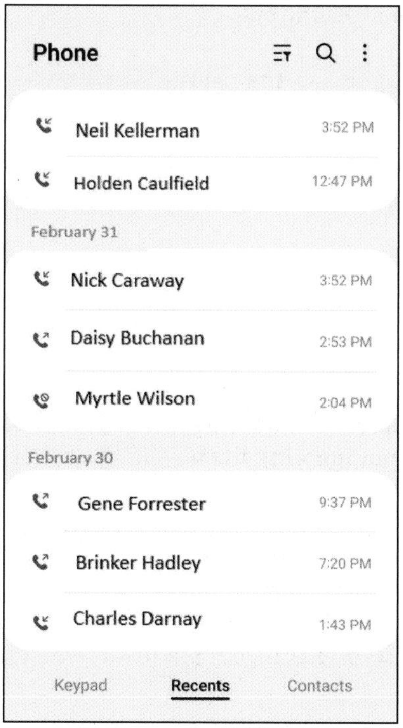

**Phone**  ≡ᵥ  Q  ⋮

| | | |
|---|---|---|
| ☎ | Neil Kellerman | 3:52 PM |
| ☎ | Holden Caulfield | 12:47 PM |

February 31

| | | |
|---|---|---|
| ☎ | Nick Caraway | 3:52 PM |
| ☎ | Daisy Buchanan | 2:53 PM |
| ☎ | Myrtle Wilson | 2:04 PM |

February 30

| | | |
|---|---|---|
| ☎ | Gene Forrester | 9:37 PM |
| ☎ | Brinker Hadley | 7:20 PM |
| ☎ | Charles Darnay | 1:43 PM |

Keypad    **Recents**    Contacts

**FIGURE 3-5:**
The Recents screen on your Phone app.

Adjacent to the number for each call is an icon meaning the following:

>> **Outgoing call you made:** An arrow pointing to the phone silhouette

>> **Incoming call you received:** An arrow pointing away from the phone silhouette

>> **Incoming call you missed:** A bent arrow above the phone silhouette

>> **Incoming call you ignored:** A blue slash sign next to the phone silhouette

The log is a list of all the calls you made or were made to you, which is handy if you want to call someone again or call them back. By tapping any name in your call list, you see a screen like the one shown in Figure 3-6. From this pop-up within the screen, you can do several things:

>> Call the number by tapping the circle with the phone silhouette.

>> Send a text to that number by tapping the talk bubble icon. (More on this in Chapter 4.)

**FIGURE 3-6:** The phone number pop-up on the Recents screen.

>> Make a video call if you're set up to do that.

>> Tap the call log details icon (the *i* in the circle) to add the number to your contacts list. (I cover contacts in more detail in Chapter 6.)

When you tap the call log detail icon in Figure 3-6, you can see the recent history of all the communication you've had with this person, as shown in Figure 3-7. This includes calls, texts, and video calls.

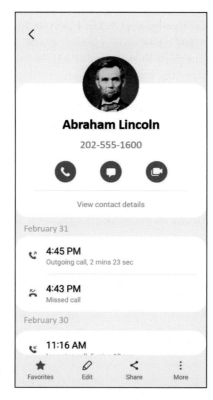

**FIGURE 3-7:**
Call Log detail.

# Making an Emergency Call: The 411 on 911

Cellphones are wonderful tools for calling for help in an emergency. The Samsung Galaxy S22, like all phones in the United States and Canada, can make emergency calls to 911.

Just tap the Phone icon on the Home screen, tap 911, and then tap Send. You'll be routed to the 911 call center nearest to your location. This works wherever you're at within the United States. So, say you live in Chicago but have a car accident in Charlotte; just tap 911 to be connected to the 911 call center in Charlotte, not Chicago.

Even if your phone isn't registered on a network, you don't have a problem as long as you have a charge in the battery. You phone lets you know that the only number you can dial is a 911 call center, even if the Home screen is locked.

TIP

When you call 911 from a landline, the address you're calling from is usually displayed for the operator. When you're calling from a cellphone, though, the operator doesn't have that specific information. So, when you call 911, the operator might say, "911. *Where* is your emergency?" Don't let this question throw you; after all, you're probably focused on *what* is happening and not on *where*. Take a moment and come up with a good description of where you are — the street you're on, the nearest cross street (if you know it), any businesses or other landmarks nearby. An operator who knows where you are is in a better position to help you with your emergency. Your phone does have a GPS receiver in it that 911 centers can access. However, it's not always accurate; it may not be receiving location information at that moment, as is the case when you're indoors.

REMEMBER

When traveling outside the United States or Canada, 911 might not be the number you call in an emergency. Mexico uses 066, 060, or 080, but most tourist areas also accept 911. And most — but not all — of Europe uses 112. Knowing the local emergency number is as important as knowing enough of the language to say you need help.

## WHEN YOU ACCIDENTALLY DIAL 911

If you accidentally dial 911 from your phone, don't hang up. Just tell the operator that it was an accidental call. They might ask some questions to verify that you are, indeed, safe and not being forced to say that your call was an accident.

If you panic and hang up after accidentally dialing 911, you'll get a call from the nearest 911 call center. Always answer the call, even if you feel foolish. If you don't answer the call, the 911 call centers will assume that you're in trouble and can't respond. They'll track you down from the GPS in your phone to verify that you're safe. If you thought you'd feel foolish explaining your mistake to a 911 operator, imagine how foolish you'd feel explaining it to the police officer who tracks you down and is upset with you for wasting the department's time.

# Syncing a Bluetooth Headset

With a Bluetooth headset device, you can talk on your phone without having to hold the phone to your ear — and without any cords running from the phone to your earpiece. You've probably come across plenty of people talking on Bluetooth headsets. You might even have wondered whether they were a little crazy talking to themselves. Well, call yourself crazy now, because when you start using a Bluetooth headset, you might never want to go back.

Not surprisingly, Galaxy S22 phones can connect to Bluetooth devices. The first step to using a Bluetooth headset with your phone is to sync the two devices. Here's how:

1. **From the Home screen on your phone, slide up to get to the Apps screen.**

   This gets you to the list of all the applications on your phone.

2. **Flick or pan to the Settings icon and tap it.**

   The Settings icon is shown here. This screen holds most of the settings that you can adjust on your phone. If you prefer, you can also bring down the notification screen and tap the gear icon or tap the menu button on the Home screen. All these actions will get you to the same place.

   Tapping on the Settings icon brings up the screen shown in Figure 3-8.

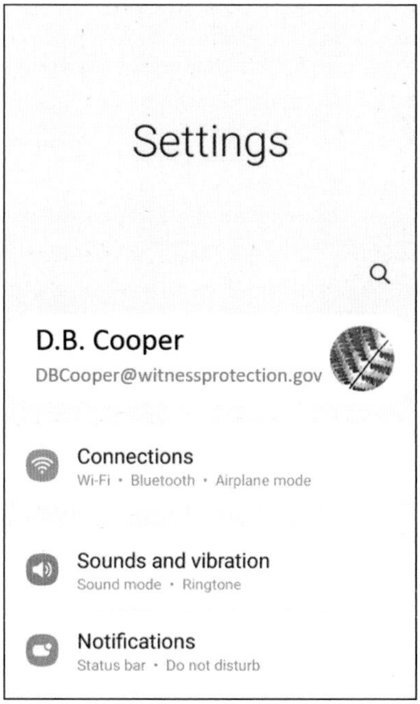

**FIGURE 3-8:** The Settings screen.

**3.** **Tap the Connections icon.**

All the options for connectivity on your phone appear.

**4.** **Tap the Bluetooth icon.**

This will bring up one of the two screens shown in Figure 3-9. If Bluetooth is off, it will looks like the screen to the left. If it is on, it will look like the screen on the right.

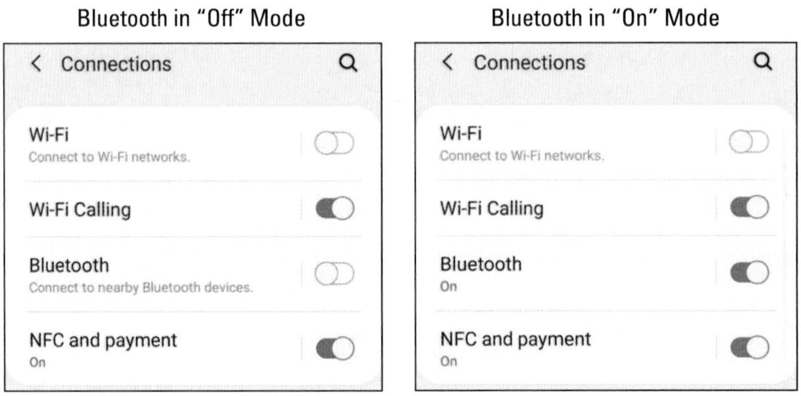

Bluetooth in "Off" Mode    Bluetooth in "On" Mode

**FIGURE 3-9:**
The Bluetooth
Settings
screens.

| < Connections     Q | | < Connections     Q |
|---|---|
| Wi-Fi<br>Connect to Wi-Fi networks. | Wi-Fi<br>Connect to Wi-Fi networks. |
| Wi-Fi Calling | Wi-Fi Calling |
| Bluetooth<br>Connect to nearby Bluetooth devices. | Bluetooth<br>On |
| NFC and payment<br>On | NFC and payment<br>On |

**5.** **Put the phone in Pairing mode by turning on Bluetooth and tapping Scan.**

This step enables your phone to be visible to other Bluetooth devices. This state will last for about 60 seconds — enough time for you to get your Bluetooth device into pairing mode so that both devices can negotiate the proper security settings and pair up every time they "see" each other from now on.

**6.** **Next, put your headset into sync mode.**

Follow the instructions that came with your headset.

After a moment, the phone "sees" the headset. When it does, you're prompted to enter the security code, and the software keyboard pops up.

**7.** **Enter the security code for your headset and then tap the Enter button.**

**TIP**

The security code on most headsets is 0000, but check the instructions that came with your headset if that number doesn't work.

Your phone might see other devices in the immediate area. If so, it asks you which device you want to pair with. Tap the name of your headset.

Your headset is now synced to your phone. If you turn one on when the other is already on, they recognize each other and automatically pair up.

**TIP**

Some manufacturers make it easier to link via Bluetooth. For example, you may have received a pair of Samsung Galaxy Buds. With these ear buds, you simply open the case where you store the buds with your phone in scan mode. You're asked to connect, and it's all done. Don't be surprised when it's even easier than the steps listed in this section!

# Options Other than Headsets

Headsets are not the only option anymore. Although many people walk around with the ubiquitous Bluetooth headset dangling from an ear, many other choices are out there. You can sync to all kinds of Bluetooth devices, including external keyboards, laptops, tablets, external speakers, and even your car.

There are also external sensors for measuring blood pressure, heart rate, and a lot of other physical information. Some of these are built into wearable devices, like a fitness bracelet. Manufacturers are now embedding computers into home appliances that can also connect to your smartphone through Bluetooth!

The good news is that regardless of the technology, you can connect all these devices to your phone simply by using the steps described in the preceding section. I talk more about these possibilities in later chapters.

Chapter **4**

# Discovering the Joy of Text

Sure, cellphones are made for talking. But these days, many people use their cellphones even more for texting.

Even the most basic phones support texting these days, but your Galaxy S22 phone makes sending and receiving text messages more convenient, no matter whether you're an occasional or pathological texter. In this chapter, I fill you in on how to send a text message (with or without an attachment), how to receive a text message, and how to read your old text messages.

**TIP**

This chapter uses images from the Samsung Messaging application. It is possible that your phone may have as its default another texting application. If so, you can easily switch to the Messaging app. You can also use the default app, but the images will be somewhat different. Your choice.

## Sending the First Text Message

Text messages (which are short messages, usually 160 characters or less, sent by cellphone) are particularly convenient when you can't talk at the moment (maybe you're in a meeting or class) or when you just have a small bit of information to share ("Running late — see you soon!").

Many cellphone users — particularly younger ones — prefer sending texts to making a phone call. They find texting a faster and more convenient way to communicate, and they often use texting shorthand to fit more content in a short message.

There are two scenarios for texting. The first is when you send someone a text for the first time. The second is when you have a text conversation with a person.

When you first get your phone and are ready to brag about your new Galaxy S22 and want to send a text to your best friend, here's how easy it is:

1. **On the Home screen, tap the Messages icon**

   The Messages icon looks like a squared-off voice bubble from a comic strip within a blue circle. When you tap it, you will get a mostly blank Home screen for texting. This is shown in Figure 4-1.

   When you have some conversations going, it begins to fill up.

**FIGURE 4-1:** The initial Messaging Home screen.

2. **Tap the New Message icon.**

   Tapping the New Message icon brings up the screen shown in Figure 4-2.

3. **Tap to enter the recipient's ten-digit mobile telephone number.**

   A text box appears at the top of the screen with the familiar To field at the top. The keyboard appears at the bottom of the screen.

   As shown in Figure 4-3, the top field is where you type the telephone number. The numerals are along the top of the keyboard.

   **REMEMBER**

   Be sure to include the area code, even if the person you're texting is local. There's no need to include a 1 before the number.

   If this is your first text, you haven't had a chance to build up a history of texts. After you've been using your messaging application for a while, you will have entered contact information, and your phone will start trying to anticipate your intended recipient. You can take one of its suggestions or you can just keep on typing.

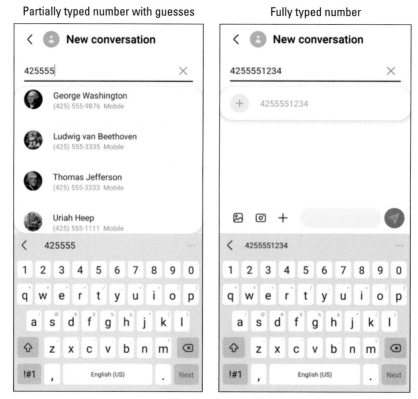

Partially typed number with guesses · Fully typed number

FIGURE 4-3:
Type the
recipient's
number in the
upper text box.

4. **To type your text message, tap the text box that says Enter Message. Figure 4-4 shows you where to enter your text.**

Your message will appear in the text box to the right of the paper clip icon.

In the Android Messaging app, your text message can be up to 160 characters, including spaces and punctuation. The application counts down the number of characters you have left.

**REMEMBER**

5. **Send the text by tapping the Send button to the right of your message.**

The Send button is not present before you start typing. After you type something, a blue-green circle with the silhouette of a paper airplane turns blue-green. Once you tap the blue-green button, the phone takes it from here. Within a few seconds, the message is sent to your friend's cellphone.

**TIP**

After you build your contact list (read about this in Chapter 6), you can tap a name from the contact list or start typing a name in the recipient text box. If there's only one number for that contact, your phone assumes that's the receiving phone you want to send a text to. If that contact has multiple numbers, it asks you which phone number you want to send your text to.

**FIGURE 4-4:**
Type your text.

**REMEMBER**

In most cases, the default for your texting app is for your phone to automatically correct what it thinks is a misspelled word. You can see it guess on the darker gray area under the text message. This capability is called *autocorrect.* You may find it very handy, or you may find it annoying. If you like it, you should still verify that it corrected the word in the right way. If you want evidence as to why this is a good idea, search "funny autocorrect examples" in your favorite search engine (although some can be very racy).

**WARNING**

You've probably heard a thousand times about how it's a very bad idea to text while you're driving. Here comes one-thousand-and-one. It's a *very bad idea* to text while you're driving — and illegal in many places. There are Dummies who read this book, who are actually very smart, and then there are DUMMIES who text and drive. I want you to be the former and not the latter.

# Carrying on a Conversation via Texting

In the bad ol' pre-Galaxy S days, most cellular phones would keep a log of your texts. The phone kept the texts that you sent or received in sequential order, regardless of who sent or received them.

Texts stored sequentially are old-school. Your Galaxy S22 keeps track of the contact with whom you've been texting and stores each set of back-and-forth messages as a *conversation*.

In Figure 4-5, you can see that the first page for messaging holds *conversations*. After you start texting someone, those texts are stored in one conversation.

As Figure 4-5 shows, each text message is presented in sequence, with the author of the text indicated by the direction of the text balloon.

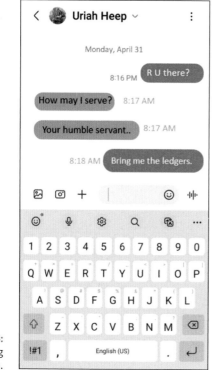

FIGURE 4-5:
A messaging
conversation.

Note the Enter Message text box at the bottom of the screen. With this convenient feature, you can send whatever you type to the person with whom you're having a conversation.

In the bad old days, it was sometimes hard to keep straight the different texting conversations you were having. When you start a texting conversation with someone else, there is a second conversation.

Before too long, you'll have multiple conversations going on. Don't worry. They aren't the kind of conversations you need to keep going constantly. No one thinks twice if you don't text for a while. The image in Figure 4-6 shows how the text page from Figure 4-1 can look before too long.

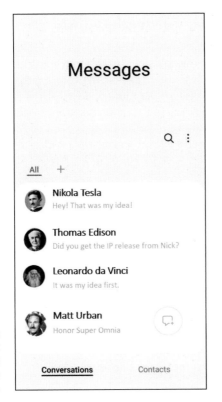

**FIGURE 4-6:**
The Text screen showing multiple conversations.

It is easy to change the font size of conversations. To make the fonts larger and easier to read, use the stretch motion. Use the pinch motion to make the fonts smaller so that you can see more of the conversation.

# Sending an Attachment with a Text

What if you want to send something in addition to or instead of text? Say you want to send a picture, some music, or a Word document along with your text. Easy as pie, as long as the phone on the receiving end can recognize the attachment. Here is the recipe:

1. **From the Home screen, tap the Messages icon.**

2. **Either tap the New Message icon and enter the number of the intended recipient or pick up on an existing conversation.**

   You'll see the Enter Message bubble from Figure 4-4. Enter the information you want like a normal text.

3. **To add an attachment, tap the arrow sign to the left of where you enter text.**

   The arrow sign brings up three icons: a silhouette of mountains, a silhouette of a camera, and a plus sign. If you want to attach a photo, tap the silhouette of the mountains. If you want to take a new picture or video, tap the camera icon. Otherwise, tap the plus sign. You see the options shown in Figure 4-7, which asks what kind of file you want to attach. Your choices include information on your location, audio files, and some others I describe in Chapters 9, 10, 12, 13, and 14 when you have some good files to share. For now, it's just good that you know you have options.

4. **Tap your choice of file type, and your phone presents you with the options that fall into that category.**

   Get ready to be amazed. Underneath the text box, you have five choices. These include the Gallery for photos, a shot you can take now with the camera, a doodle, a map with your location, or other categories. Figure 4-7 shows the Other category. This is for any file you want to attach that is already on your phone. It can be your electronic business card. It could be a video.

   The Quick response opens up canned responses, such as "What's up?" or "Where's the meeting?" How cool is that?!

5. **Continue typing your text message, if needed.**

6. **When you're done with the text portion of the message, tap the Send button (the one that looks like a line drawing of a paper airplane), and off it all goes.**

**TECHNICAL STUFF**

A simple text message is an SMS (short message service) message. When you add an attachment, you're sending an MMS (multimedia messaging service) message. Back in the day, MMS messages cost more to send and receive than SMS messages did. These days, that isn't the case in the United States.

# Receiving Text Messages

Receiving a text is even easier than sending one.

When you're having a text conversation and you get a new text from the person you're texting with, your phone beeps and/or vibrates. Also, the notification area of the screen (the very top) shows a very small version of the Messages icon.

You can either pull down the notification area from the very top of the screen or start the messaging application. Your choice.

If an attachment comes along, it's included in the conversation screen.

# Managing Your Text History

The Messaging Conversations screen stores and organizes all your texts until you delete them. You should clean up this screen every now and then.

The simplest option for managing your messages is to tap the Menu icon and then tap Delete. You can then select and unselect all the conversations that you want deleted. Tap the icon that looks like a garbage can at the bottom of the screen, and they disappear.

Another deletion option is to open the conversation. You can delete each text by pressing and holding on the balloon. After a moment, a menu appears from which you can delete that message. This method is a lot slower if you have lots of texts, though.

I recommend that you be vicious in deleting the older texts and conversations. Trust me; deleting all your old messages can be cathartic!

# Chapter **5**

# Sending and Receiving Email

I f you've had email on your phone for a while, you know how convenient it is. If your Galaxy S22 phone is your first cellphone with the capability to send and receive email, prepare to be hooked.

I start this chapter by showing you how to set up your email, regardless of whether your email program is supported. Then I show you how to read and manage your email. Finally, I tell you how to write and send email in the last section in this chapter.

## Setting Up Your Email

These days, many people have multiple personal email addresses for many reasons. Your phone's Email app can manage up to ten email accounts. With a Galaxy S22 phone, you'll want to create Google account if you don't already have one. If you don't have a Google account, you'll miss out on so many exciting capabilities that it's almost worth settling for a lesser phone.

If you have an email account that ends in @gmail.com, by default you have a Google account. If you do *not* have an account that ends in @gmail.com, you can either create a Google account with your existing email or create a separate email account on Google's Gmail just for your phone. Setting up a new Gmail account if you don't have one already (see the upcoming section "Setting up a new Gmail account") is easy, but it isn't too hard to use an existing email for a Google account.

The Email app on your phone routinely "polls" all the email systems for which you identify an email account and password. It then presents you with copies of your email.

TIP

Your phone mainly interacts with your inbox on your email account. It isn't really set up to work like the full-featured email application on your computer, though. For example, many email packages integrate with a sophisticated word processor, have sophisticated filing systems for your saved messages, and offer lots of fonts. As long as you don't mind working without these capabilities, you might never need to get on your computer to access your email again, and you could store email in folders on your phone.

Setup is easy, and having access to all your email makes you so productive that I advise you to consider adding all your email accounts to your phone.

## Getting ready

In general, connecting to a personal email account simply involves entering the name of your email account(s) and its password(s) in your phone. Have these handy when you're ready to set up your phone.

As mentioned, you can have up to ten email accounts on your phone; however, you do need to pick one account as your favorite. You can send an email using any of the accounts, but your phone wants to know the email account that you want it to use as a default.

Next, you may want to have access to your work account. This is relatively common these days, but some companies see this as a security problem. You should consult with your IT department for some extra information. Technologically, it's not hard to make this kind of thing happen as long as your business email is reasonably modern.

Finally, if you don't already have a Gmail account, I strongly encourage you to get one. Read the nearby sidebar, "The advantages of setting up a Google account" to find out why.

## Setting up your existing Gmail account

If you already have a Gmail account, setting it up on your phone is easy as can be. Follow these steps from the Apps menu:

1. **Find the Gmail icon in the Apps list.**

   Here is a confusing part. The icon on the left in Figure 5-1 is the Gmail app. The icon on the right is for the general email app. The general email app is for your combined email accounts. The general email account is the app that you will use to access any and all of your email accounts.

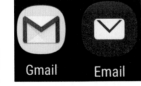

**FIGURE 5-1:**
The Email and
Gmail icons.

2. **Tap the Gmail icon.**

   Because your phone does not know if you have a Gmail account yet, it offers you the option of entering your Gmail account or whether you want to create a new account. This page is shown in Figure 5-2.

**FIGURE 5-2:**
Add your
account page.

3.  **Enter your Gmail account email address and tap Next.**

    Be sure to include the @gmail.com suffix.

4.  **Enter your existing Gmail password and tap Next.**

    Go ahead and type your password. When you're ready, tap Next on the keyboard.

You may get a pop-up reconfirming that you agree with the terms of use and all that legal stuff. Tap OK. You'll see lots of flashing lights and whirling circles while your phone and your Gmail account get to know each other.

If everything is correct, your phone and your account get acquainted and become best friends. After a few minutes, they are ready to serve your needs. There are even a few screens that tell you all the wonderful things that your Gmail account will do for you. Believe every word!

If you have a problem, you probably mistyped something. Try retyping your information. From this point on, any email you get in your Gmail account will also appear on your phone!

# Setting up a new Gmail account

If you need to set up a new Gmail account, you have a few more steps to follow. Before you get into the steps, think up a good user ID and password.

TIP

Gmail has been around for a while. That means all the good, simple email addresses are taken. Unless you plan to start using this email account as your main email, which you could do if you wanted, you're probably best off if you pick some unusual combination of letters and numbers that you can remember for now to get through this process.

When you have all of this ready, follow Steps 1 and 2 in the previous section, but tap the Create Account link when you get to the screen shown in Figure 5-2. From there, follow these steps:

1. **Enter your first and last names on the screen.**

   Google asks for your name in the screen shown in Figure 5-3. This is how Google personalizes any communications it has with you.

TIP

   You may be tempted to use a fake name or some other clever two-word combination in place of a name. Don't do it. You will still be getting email to Rita Book or Warren Peace long after the humor has worn off.

2. **Enter the username you want to use with Gmail.**

   On the screen shown in Figure 5-3, enter the username you want. Hopefully you get this name approved on the first shot.

   If your first choice isn't available, try again. There is no easy way to check before you go through these steps. Eventually, you hit on an unused ID, or you will use one of the suggestions in blue font. When you're successful, it will congratulate you.

3. **Pick a good password of at least eight characters.**

   Be sure to pick one that is unique and that you can remember.

4. **Enter the other information and an alternate email address and tap Continue.**

   If you forget your password, Google wants to verify that you're really you and not someone pretending to be you. Google does this check by sending a confirmation email to another account or sending a text to your phone. You can do this now or tap Skip to do it later.

After you tap Done, light flashes, and you see the screen working away. This process usually takes less than two minutes. While you wait, you'll see all kinds of messages that it's getting ready to sync things. Don't worry. I explain these messages in good time. For now, you can add any other email accounts you want by following the steps in the next section.

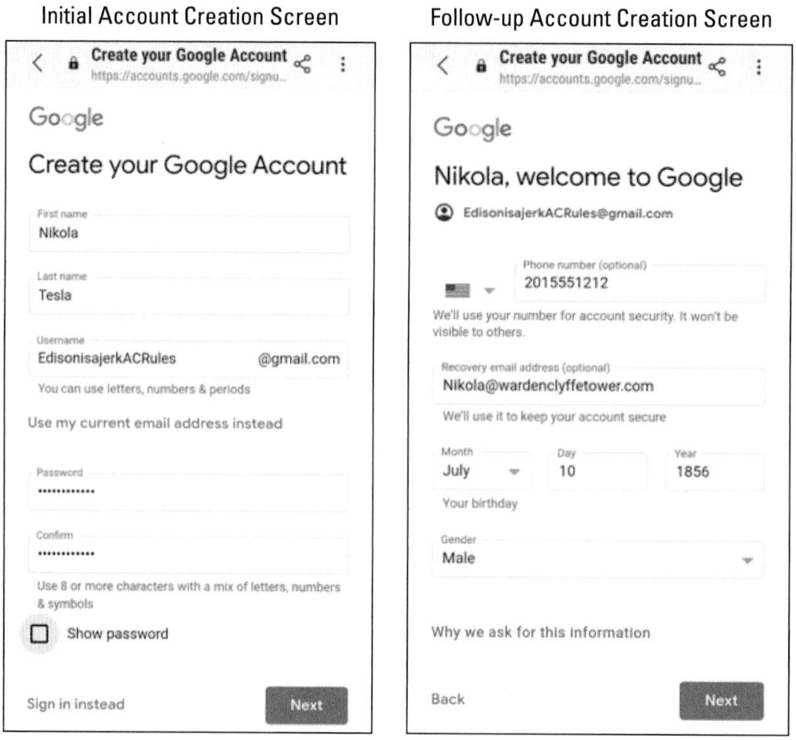

**FIGURE 5-3:**
The Create
Your Google
Account
screen.

# Working with non-Gmail email accounts

Your phone is set up to work with up to ten email accounts. If you have more than ten accounts, I'm thinking that you might have too much going on in your life. No phone, not even the Galaxy S22, can help you there!

I now cover the steps to adding your email accounts. Once you tell your phone all your email accounts, the Email screen will let you see the inbox of each account or combine all your email in a single inbox. You can choose which option works best for you.

## Adding your first email account

To get started, have your email addresses and passwords ready. When you have them, go to your phone's Home screen. Look for the Mail icon; it is an envelope in silhouette (refer to Figure 5-1). This is probably on your Home screen as one of the four primary shortcuts just above the Device Function keys or in your application list.

1. **Tap the Mail icon from the Home screen.**

   This brings up a screen like the one shown in Figure 5-4. Choose one of the options to get started. In every case, it starts with entering your email address.

**WARNING**

   Your email address should include the full shebang, including the @ sign and everything that follows it. Make sure to enter your password correctly, being careful with capitalization if your email server is case-sensitive (most are). If in doubt, select the option that lets you see your password.

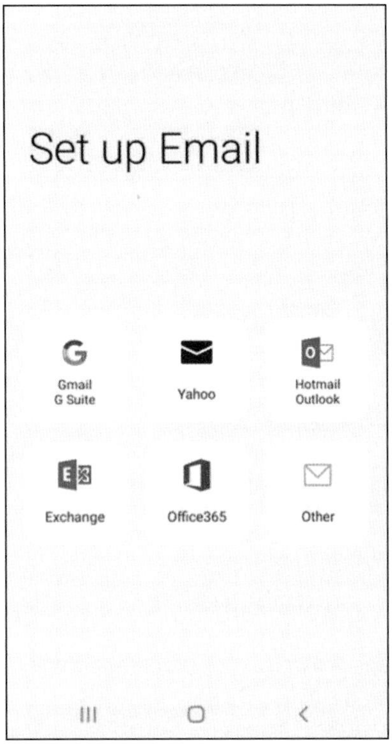

## Set up Email

G
Gmail
G Suite

✉
Yahoo

O
Hotmail
Outlook

E
Exchange

Office365

✉
Other

**FIGURE 5-4:**
The setup screen for the Email app.

2. **Tap the box to enter your password.**

   Don't be surprised if your S22 brings you to a special screen specifically for your email provider for you to enter your password. Figure 5-5 shows the sign-in screen for a generic email account. Figure 5-6 shows the screen for a Microsoft Exchange account. One of these will work.

**FIGURE 5-5:**
The sign-in
screen for a
generic email
account.

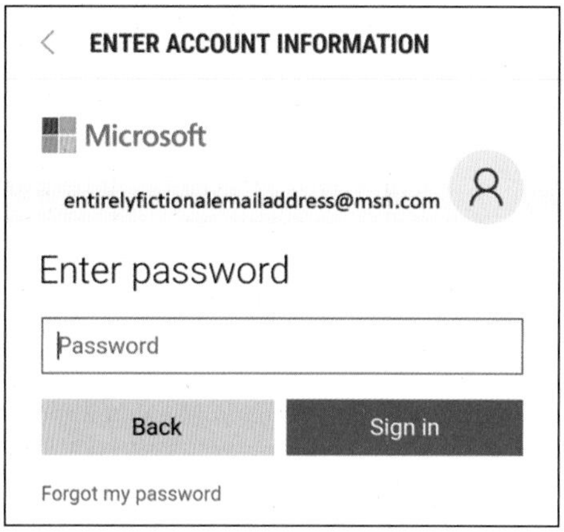

> **< Add new account**
>
> Email address
>
> Password
>
> ○ Show password
>
> Manual setup                    Sign In

**FIGURE 5-6:**
The Enter
Account
Information
screen for
a Microsoft
Exchange
account.

> **< ENTER ACCOUNT INFORMATION**
>
> ■ Microsoft
>
> entirelyfictionalemailaddress@msn.com  ⊙
>
> Enter password
>
> Password
>
> [ Back ]   [ Sign in ]
>
> Forgot my password

**3. Carefully enter your password in the appropriate field and tap Next.**

You're asked for permissions. Just go with the default settings for now. The screen shown on the left in Figure 5-7 appears.

**4. Tap the Sync account link.**

The screen shown on the right in Figure 5-7 appears.

**5. Select your desired Sync Settings and then tap Next.**

You can select how often you want your phone and the email service to synchronize. A lot of thought and consideration has been put into the default settings. If you just want to get started, tap Next.

Initial Account Screen          Sync Account Screen

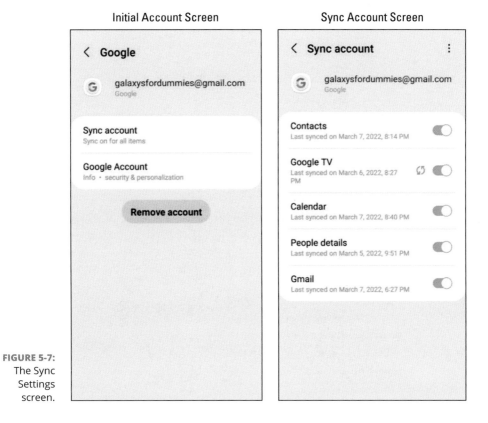

**FIGURE 5-7:**
The Sync
Settings
screen.

If you want to fine-tune things later, it is not hard to go back and adjust these settings. These settings are intended to be gentle on your data usage.

If you want images in email to download immediately, store older email messages on your phone, or check to see whether you have new email all the time, you can change these settings on this page for this email account. If you know what you want that is different from the default settings, make the changes and then tap Next.

**6.** **Tap Done.**

Using Figure 5-8 as an example, you can see that my account is now registered on my phone. It worked!

Options icon

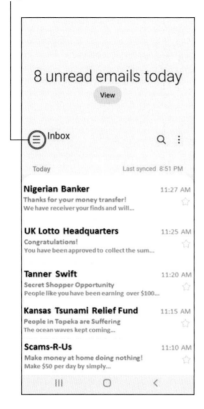

**FIGURE 5-8:**
The Email
Home screen.

## Adding additional email accounts

Once you have entered your first email account, there are a few different steps to add additional accounts.

1. **Tap the Options icon at the top-left part of the screen.**

   The three horizontal lines bring up the slide-in screen shown in Figure 5-9. This allows you to see other email folders. It also lets you access the setting icon.

2. **Tap the Settings icon.**

   Tapping Settings brings up the screen shown in Figure 5-10.

3. **Tap Add Account next to the green plus sign.**

   This brings you back to the email setup screen (refer to Figure 5-5).

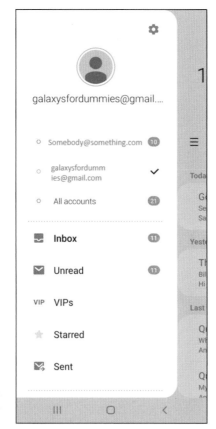

**FIGURE 5-9:**
The Options
slide-in screen.

At this point you can add up to nine more accounts, and remember, you will be asked which email account you want to be your primary account. It is entirely up to you. You can send and receive email from all your accounts by selecting it, but only one can be the primary account used if you send an email from another application, such as the Contacts app.

## Setting up a corporate email account

In addition to personal email accounts, you can add your work email account to your phone — if it's based upon a Microsoft Exchange server, that is, and if it's okay with your company's IT department.

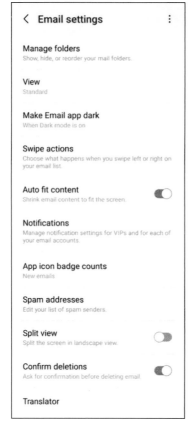

**FIGURE 5-10:**
The Email
Settings
screen.

Before you get started, you need some information from the IT department of your company:

>> The domain name of the office email server

>> Your work email password

>> The name of your exchange server

If the folks in IT are okay with you using your phone to access its email service, your IT department will have no trouble supplying you with this information.

**WARNING**

Before you set up your work email on your phone, make sure that you have permission. If you do this without the green light from your company, and you end up violating your company's rules, you could be in hot water. Increasing your productivity won't be much help if you're standing out in the parking lot holding all the contents of your office in a cardboard box.

Assuming that your company wants you to be more productive with no extra cost to the company, the process for adding your work email starts at your email Home screen shown in Figure 5-5.

1. **Enter your email address and password.**

   The Corporate screen is shown in Figure 5-11.

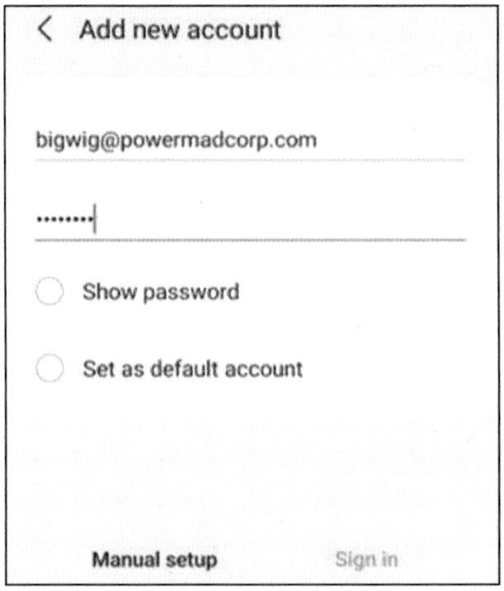

**FIGURE 5-11:**
The Corporate setup email screen.

2. **Tap Manual Setup.**

   This brings up the screen shown in Figure 5-12.

   Chances are that you don't have the foggiest notion what any of this means or what you are to do now.

3. **Verify that your IT department is good with you having email on your own device and have them give you the necessary settings.**

Seriously. It is increasingly common that a firm will give you access on your phone. At the same time, companies do not generally circulate documents with how to make this happen. This would be a big security problem if anyone could get access. Save yourself the time and get help from IT.

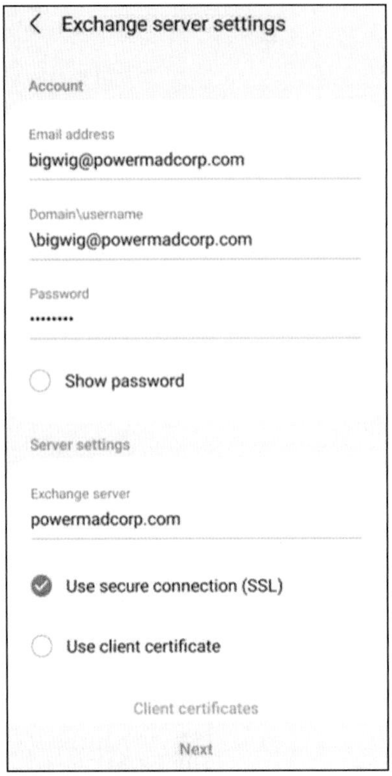

**FIGURE 5-12:**
The Manual
Setup screen
for adding
corporate
email
accounts.

# Reading Email on Your Phone

In Figure 5-8, you can see how the email screen looks for my Gmail account. You can also set up this screen so that it combines all your email into one inbox. At any given time, you might want to look at the accounts individually or all together.

To look at all your email in one large inbox, tap the options icon to bring up the screen in Figure 5-9, and select All Accounts. This lists all your email in chronological order. To open any email message, just tap it.

If, on the other hand, you want to see just email from one account, tap the options icon, and then tap the account you want to focus on at the moment. Your phone brings up your email in chronological order for just that email address.

# Writing and Sending Email

After you set up the receiving part of email, the other important side is composing and sending email. At any time when you're in an email screen, simply tap the Menu button to get a pop-up screen. From the pop-up menu, tap the Compose icon to open the email composition screen.

Here's the logic that determines which email account will send this email:

>> If you're in the inbox of an email account and you tap the Compose icon after tapping Menu, your phone sends the email to the intended recipient(s) through that account.

>> If you're in the combined inbox or some other part of the email app, your phone assumes that you want to send the email from the default email account that you selected when you registered your second (or additional) email account.

When you tap the Compose icon in the Menu pop-up menu, it tells you which account it will use. The Email composition screen shown in Figure 5-13 says this email will be coming from this account: galaxysfordummies@gmail.com.

As shown in this screen, the top has a stalwart To field, where you type the recipient's address. You can also call up your contacts, a group, or your most recent email addresses. (Read all about contacts in Chapter 6.) Tap the address or contact you want, and it populates the To field.

Below that, in the Subject field, is where you enter the email's topic. And below that is the body of the message, with the default signature, *Sent from my Samsung Galaxy S22*, although your cellular carrier might have customized this signature.

At the top of the screen are three links:

>> **Attach:** Tap this hyperlink to attach a file of any variety to your email.

>> **Send:** Tap this icon that looks like a paper airplane to send the email to the intended recipient(s).

**FIGURE 5-13:**
The Email
composition
screen.

>> **More (three vertical dots):** Tap this option, and you get the pop-up shown in Figure 5-14. This may give you some of the following options:

- **Save in Drafts:** This allows you to complete the email later without losing your work.

- **Send email to myself:** This is an alternative to saving the email in your Sent mail folder.

- **Priority:** This option signifies that this is more important that the average email to your recipients.

- **Security options:** You have some fancy, 007 options here. You can sign the document or encrypt it.

- **Turn on Rich text:** To save data usage, email is sent in a default font. If you want to burn at little more data, you can get creative and add more elaborate fonts.

- **Permission:** This is an advanced setting to have a fancy template for the background in your email.

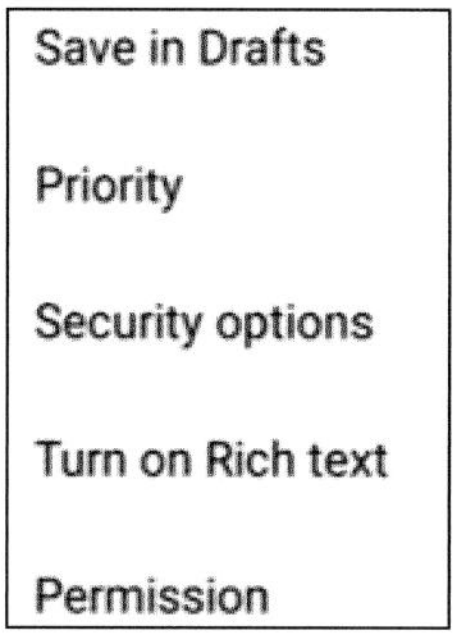

**FIGURE 5-14:**
Composition
email options.

If you change your mind about sending an email message, you can just tap the Back key. If you're partially done with the message, you're asked whether you want to save it in your Drafts folder.

The Drafts folder works like the Drafts folder in your computer's email program. When you want to continue working on a saved email, you open the Drafts folder, tap on it, and continue working.

# Replying to and Forwarding Email

Replying to or forwarding the email that you get is a common activity. You can do this from your Email app. Figure 5-15 shows a typical open email message.

You can reply by tapping the icon with the return arrow at the bottom of the screen. If other people were copied on the email, you could tap the double return arrow to Reply All.

When you tap either of these options, the Reply screen comes back with the To line populated by the sender's email address (or addresses) and a blank space where you can leave your comments. (In the case of Figure 5-15, you could ask that your dad not send you any more jokes like these.)

To forward the email, tap the arrow pointing to the right above the word Forward and enter the addressee just as you do when sending a new email.

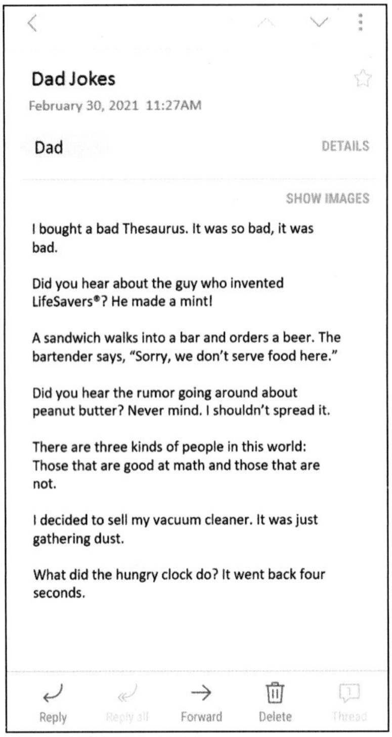

**FIGURE 5-15:**
An opened email.

Dad Jokes

February 30, 2021  11:27AM

Dad                                    DETAILS

                                    SHOW IMAGES

I bought a bad Thesaurus. It was so bad, it was bad.

Did you hear about the guy who invented LifeSavers®? He made a mint!

A sandwich walks into a bar and orders a beer. The bartender says, "Sorry, we don't serve food here."

Did you hear the rumor going around about peanut butter? Never mind. I shouldn't spread it.

There are three kinds of people in this world: Those that are good at math and those that are not.

I decided to sell my vacuum cleaner. It was just gathering dust.

What did the hungry clock do? It went back four seconds.

Reply    Reply all    Forward    Delete    Thread

» Getting all your contacts in one location

» Keeping up to date with just a few taps

# Chapter **6**

# Managing Your Contacts

You're probably familiar with using contact databases. Many cellphones automatically create one, or at least prompt you to create one. You also probably have a file of contacts on your work computer, made up of work email addresses and telephone numbers. And if you have a personal email account, you probably have a contact database of email accounts of friends and family members. If you're kickin' it old school, you might even keep a paper address book with names, addresses, and telephone numbers.

The problem with having all these contact databases is that it's rarely ever as neat and tidy as I've just outlined. A friend might email you at work, so you have them in both your contact databases. Then their email address might change, and you update that information in your personal address book but not in your work one. Before long, you have duplicate and out-of-date contacts, and it's hard to tell which is correct. How you include Facebook or LinkedIn messaging in your contact profile is unclear.

In addition to problems keeping all your contact databases current, it can be a hassle to migrate the database from your old phone. Some cellular carriers or firms have offered a service that converts your existing files to your new phone, but it's rarely a truly satisfying experience. You end up spending a lot of time correcting the assumptions it makes.

You now face that dilemma again with your Galaxy S22: deciding how to manage your contacts. This chapter gives you the advantages of each approach so that you can decide which one will work best for you. That way, you won't have the frustration of wishing you had done it another way before you put 500 of your best friends in the wrong filing system.

# Using the Galaxy S22 Contacts App

Your phone wants you to be able to communicate with all the people you would ever want to in any way you know how to talk to them. This is a tall order, and your Galaxy S22 makes it as easy as possible. In fact, I wouldn't be surprised if the technology implemented in the Contacts app becomes one of your favorite capabilities of the phone. After all, your phone is there to simplify communication with friends, family, and coworkers, and the Contacts app on your phone makes it as easy as technology allows.

At the same time, this information is only as good as your contact database discipline. The focus of this section is to help you help your phone help you.

## Learning the Contacts app on your phone

The fact of the matter is, if you introduced your phone to your email accounts in Chapter 5, the Contacts list on your phone has all the contacts from each of your Contacts lists.

Take a look at it and see. From your Home screen, tap the Contacts icon.

If you haven't created a Gmail account, synced your personal email, or created a contact when you sent a text or made a call, your Contacts list will be empty. Otherwise, you see a bunch of your contacts now residing on your phone (as shown in Figure 6-1).

The top of the Contacts list includes the contacts with which you most frequently interact. As you swipe upward, the rest of your contacts are in alphabetical order.

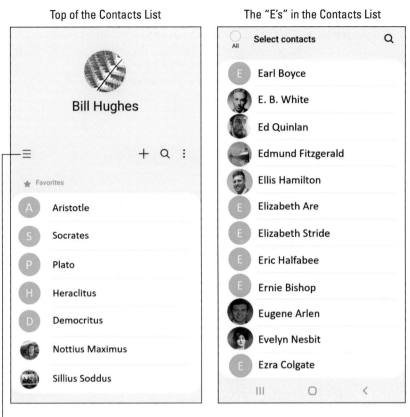

Top of the Contacts List

The "E's" in the Contacts List

**FIGURE 6-1:**
The Contacts
list.

Menu

This Contacts database does more than just store names, phone numbers, and email addresses. However, you can have the contact include any or all of the following information:

» All telephone numbers, including

- Mobile

- Home

- Work

- Other

» Email addresses

- Home

- Work

- Other

- » Instant Messaging addresses
  - WhatsApp
  - Facebook
  - Hangouts
  - QQ
  - Skype
  - Yahoo!
  - AIM
  - ICQ
  - Jabber
  - Windows Live
  - Other
- » Company/organization
- » Job title
- » Nickname
- » Mailing address for
  - Home
  - Work
  - Another location
- » Any notes about this person
  - Web address
  - Important dates
  - Relationships
  - QQ
  - A phonetic spelling of the name

As if all this weren't enough, you can assign a specific ringtone to play when a particular person contacts you. I cover the steps to assign a music file to an individual caller in Chapter 12.

Finally, you can assign a picture for the contact. It can be one out of your Gallery; you can take a new picture; or you can connect to a social network like Facebook and use that person's profile picture.

Fortunately, the only essential information is a name. Every other field is optional and is only displayed if the field contains information to be displayed. Figure 6-2 shows a lightly populated contact.

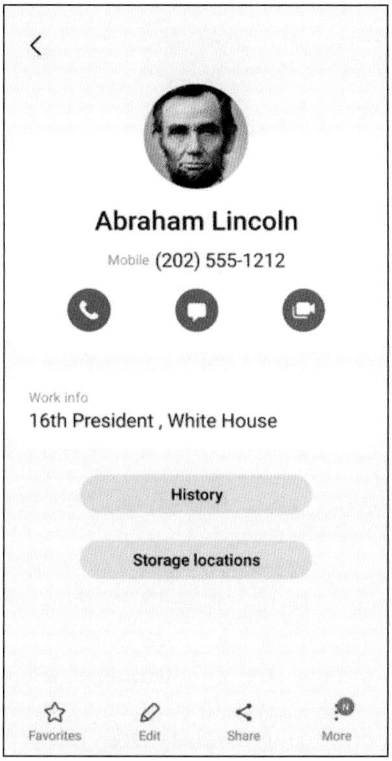

**FIGURE 6-2:**
A lightly populated contact.

## Deciding where to store your contacts

Before you get too far, it is important for you to decide where you want to store your contacts. Deciding this now with the information this section provides you will make your life much easier as you go forward. It is possible to combine contact databases, but even the best tools are imperfect.

The first time you try to save a new contact, the Contacts App will offer you a pop-up screen, as shown in Figure 6-3. This lists the options you have and will take your first entry as the option you want as a default.

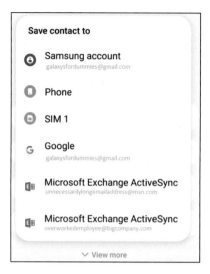

**FIGURE 6-3:**
Choosing a default for creating new contacts.

Whenever you save a new entry, you can manually switch to another of these databases. Save yourself some time and aggravation and decide now what you want to do. Here are your options on where to save new contacts:

>> Within your Samsung account

>> Within the memory of your phone

>> On the SIM card inserted in your phone

>> Within your Gmail account

>> As a contact in one of your other accounts

Each of these options has advantages and disadvantages. If you simply want my advice, I suggest using your Gmail account to store new contacts. If this satisfies you, skip to the next section on linking contacts from other sources. Read on if you need some context.

Here is the deal: If everything is working properly, all options work equally well. However, if you switch phones regularly, ever lose you phone, need to make significant updates to your contacts, or are frequently out of wireless coverage, such as on an airplane, you should consider the options.

The first option mentioned involves storing your contacts within your phone. This is a great option, as long as you have your phone. However, there will come a day when Samsung (or HTC, or LG, and so on) has something faster and better, and you will want to upgrade. At that point, perhaps next year or in ten years, you will need to move your contacts to another location if you want to keep them. Plus, if you lose your phone, I sure hope that you have recently used one of the backup options I explore in this chapter.

The second option is to store your new contacts on your SIM card. The SIM card is familiar technology if your previous phone worked with AT&T or T-Mobile. Figure 6-4 shows a profile of a typical SIM card, next to a dime for scale, although yours probably has the logo of your cellular carrier nicely printed on the card. To the right of the SIM card is the newer micro SIM card. This is the same idea, but in a smaller package.

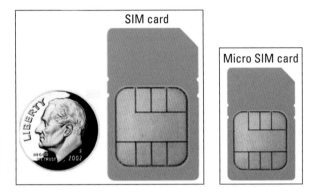

**FIGURE 6-4:**
A SIM card and a micro SIM card.

The cool aspect of using your SIM card is that you can pluck it out of your existing phone and pop it in another phone and all your contacts come with you. It works this easily if you stay with an Android smartphone of recent vintage. It almost works this easily if you, say, switch back to a feature phone (as in the kind of phone that merely makes calls and texts and costs $1). Another advantage is that your SIM card does not rely on having a wireless connection to update changes.

The next option is to store new contacts on your Gmail account. What this means is that your new contacts are automatically copied from your phone to the Gmail account of Google servers. This keeps the records on your phone, but Google maintains a copy of this record in your Gmail account. You can make changes to a contact on your phone or on your PC.

The two reasons I recommend using your Gmail account are:

>> You can maintain these records with your full-sized keyboard rather than the smaller keyboard on your phone.

>> It is easy to get back all these contacts on a new Android phone simply by telling the new phone your Gmail account. Because you will probably enter your Gmail account right away when you get a new phone, your contacts reappear quicker than they would with the fourth option.

The fourth option is to store the contacts in one of your existing email accounts. This may be the best option if you already consider your email, either your personal or work account, to be the primary location where you store contacts. If you already have good database discipline with one of your email accounts, by all means, use this email account as the default place to store new contacts.

If this particular account you currently use for keeping your contacts is not set up yet on your phone, you can tap the Add New Account process described in Chapter 5 to set up a new email account.

## Linking Contacts on your phone

The Contacts list is smart. Allow me to explain some of the things that are going on.

Say your best friend is Bill McCarty, affectionately known as "the Kid." You sent Bill a text earlier to let him know about your new phone. You followed the instructions on how to send a text in Chapter 4 and entered his telephone number. You took it to the next step and tapped Add Contact. When you were prompted to add his name, you did. Now your phone has a contact, "Bill McCarty."

Then you linked your email. Of course your buddy Bill is in your email Contacts list. So while you were reading this chapter, several things happened. First, these two contacts automatically synced. Your phone thinks about it and figures this must be the same person. It automatically combines all the information in one entry on your phone!

Then your phone automatically updates your Gmail account. On the left side of Figure 6-5, you see Bill's mobile number. This contact is synced with your Gmail account. You didn't have to do anything to make this happen.

Your phone noticed that Bill's work number was in your email contact information, but the mobile phone number you used to text him was not. No problem! Now the contact on your phone includes both the information you had in your email contact as well as his cellular phone.

Now, as slick as this system is, it isn't perfect. In this scenario, both contacts have the same first and last name. However, if the same person also goes by a different name, you have to link these contacts. For example, if you created a contact for Bill McCarty, but your email refers to him as William H. McCarty, your phone will assume that these are two different people.

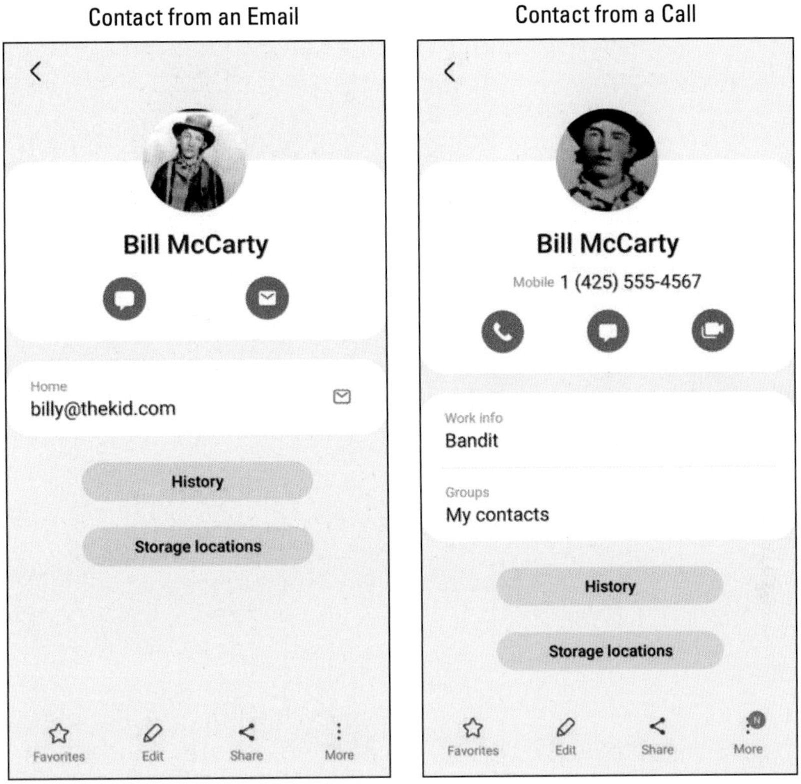

Contact from an Email                 Contact from a Call

**Bill McCarty**

**Bill McCarty**

Mobile 1 (425) 555-4567

Home
billy@thekid.com

History

Storage locations

Work info
Bandit

Groups
My contacts

History

Storage locations

☆              🖉              <              ⋮
Favorites      Edit          Share          More

☆              🖉              <              🔵
Favorites      Edit          Share          More

**FIGURE 6-5:**
Two contacts for the same person.

No problem, though. Do you see the three horizontal lines in Figure 6-1? These are the menu link. Here are the steps to link the two contacts for the same person:

**1.** **From the contact home screen, tap on the three horizontal lines.**

The pop-up shown in Figure 6-6 appears.

**2.** **From the Menu screen, tap the Manage contacts link.**

The pop-up shown in Figure 6-7 appears.

**3.** **From the Manage Contacts screen, tap the Merge Contacts link.**

The screen shown in Figure 6-8 appears.

Your phone tries to help you with some suggestions. If it gets it all wrong, you can just find the other contact by searching alphabetically. The Merge Contacts screen in Figure 6-8 is shows the contacts that have the same name, but it can also find contacts with the same number or email.

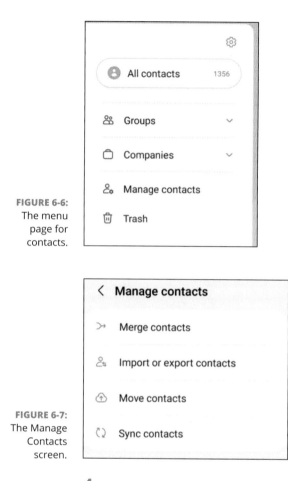

**FIGURE 6-6:**
The menu page for contacts.

**FIGURE 6-7:**
The Manage Contacts screen.

4. **Tap the contact you want joined.**

   In this case, the two contacts with the name Bill McCarty are the ones I want. Tap the button to the left of the name.

5. **Now tap the Merge link at the bottom center of the screen.**

   The combined link has all the information on this one person.

TIP

Tap to select

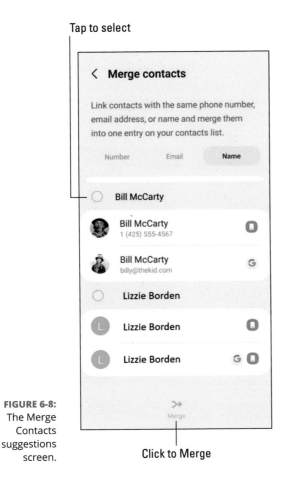

< **Merge contacts**

Link contacts with the same phone number, email address, or name and merge them into one entry on your contacts list.

| Number | Email | **Name** |

○ **Bill McCarty**

Bill McCarty
1 (425) 555-4567

Bill McCarty
billy@thekid.com

○ **Lizzie Borden**

Lizzie Borden

Lizzie Borden

Merge

**FIGURE 6-8:**
The Merge Contacts suggestions screen.

Click to Merge

# Creating Contacts within Your Database

Your phone is out there trying to make itself the ultimate contact database with as little effort on your part as possible. The interesting thing is that the salesperson in the cellular store probably didn't explain this to you in detail. It's a subtle-but-important capability that's hard to communicate on the sales floor. Here's what happens.

Whenever you make or receive a call, send or receive an email, or send or receive a text, your phone looks up the telephone number or email address from which the message originated to check whether it has seen that address before. If it has, it has all the other information on that person ready. If it doesn't recognize the originating telephone number or email, it asks whether you want to make it a new contact or update an existing one. What could be easier?

# Adding contacts as you dial

When you get a call, a text, or an email from someone who isn't in your contacts, you're given the option to create a profile for that person. The same is true when you initiate contact with someone who isn't in your Contacts list. Chapter 3 shows you the empty and full dialing screens. The image on the left in Figure 6-9 shows the phone trying its best to anticipate the person you are in the process of calling, and, if it is not saved already, the image on the right shows how it offers you the option to add this number to your contacts.

**FIGURE 6-9:** The dialing screens in process and when there is a new number.

You're immediately given the option to create a contact or update an existing contact. Your phone doesn't know whether this is a new number for an existing contact or a totally new person. Rather than make an assumption (as lesser phones on the market would do), your phone offers you the options to create a new profile or add this contact information to an existing profile.

**TIP**

Keep in mind that if you are calling an existing contact and your phone guesses the right person, you can save yourself time and tap the phone to dial.

Say that you want to enter this number as a new contact. As soon as you end the call, you see the screen on the left in Figure 6-10. Follow these steps:

1. **Tap the Add to Contacts link.**

   The screen changes to the one on the right in Figure 6-10, with a pop-up page asking if you want to create a new contact or if this number should update an existing contact.

2. **Tap the Created a New Contact link.**

   A contacts screen with the number is already populated.

3. **Fill in the correct name plus any other information that you want associated with this person.**

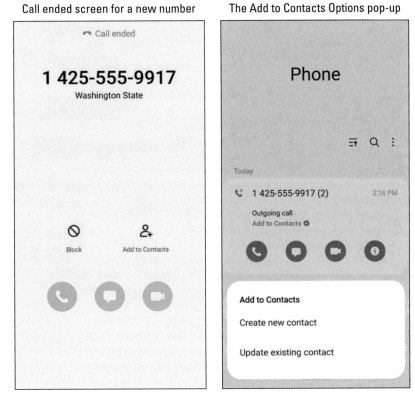

Call ended screen for a new number          The Add to Contacts Options pop-up

FIGURE 6-10:
Options when
your call is with
a new number.

# Adding contacts manually

Adding contacts manually involves taking an existing contact database and adding its entries to your phone, one profile at a time. (This option, a last resort, was the only option for phones back in the day.)

1. **Tap the Contacts icon.**

   Doing so brings up the list of existing contacts shown in Figure 6-1.

2. **Tap the Add Contacts icon (the + [plus] sign next to a silhouette).**

   This icon is shown in Figure 6-11. Tapping it will bring up a blank contact page with nothing populated.

3. **Fill in the information you want to include.**

4. **When you're done entering data, tap Save at the bottom of the screen.**

   The profile is now saved. Repeat the process for as many profiles as you want to create.

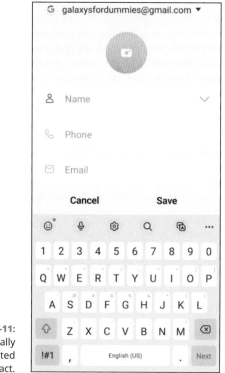

FIGURE 6-11:
A partially
populated
contact.

# How Contacts Make Life Easy

Phew. Heavy lifting over. After you populate the profiles of dozens or hundreds of contacts, you're rewarded with a great deal of convenience.

To contact someone from the Contacts directory, start by tapping the contact icon. Your Contacts list appears (refer to Figure 6-1). Scroll down or search for the contact you want. When you tap that contact name, you're given some choices, as shown in Figure 6-12.

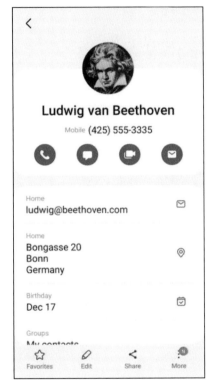

**FIGURE 6-12:**
The Contacts database when you tap a profile.

I've tapped on the contact for my friend, Ludwig, and I have choices:

>> Tap the telephone icon to dial that number.

>> Tap the message icon to send a text.

>> Tap the video call icon to have a video chat.

>> Tap the envelope icon to send an email.

About the only thing that tapping on a data field won't do is print an envelope for you if you tap the mailing address! At the bottom of the contact is a button that says History. This is a running log of all your communications with that contact regardless of the media you used. Figure 6-13 shows an example.

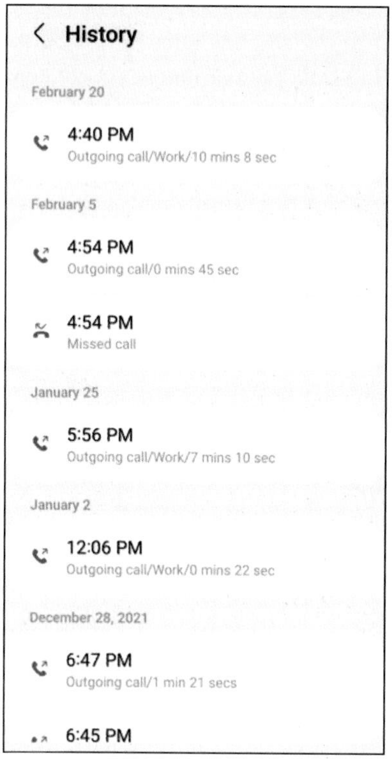

**FIGURE 6-13:**
Your history
with that
person.

It is all there.

# Playing Favorites

Over the course of time, you'll probably find yourself calling some people more than others. It would be nice to not have to scroll through your entire Contacts database to find those certain people. Contacts allow you to place some of your contacts into a Favorites list for easy access.

From within the Contacts app, open the profile and notice the star at the bottom of the screen (refer to Figure 6-12). If that star is filled in, that contact is a Favorite. If not, then not.

To make a contact into a star, tap the blank outline of the star. To demote a contact from stardom without deleting, tap the gold star.

You won't immediately see a difference in your contacts other than the appearance of the solid star. When you open your phone, however, this contact now appears under your Favorites tab. The Favorites tab is like a mini-contact database. It looks similar in structure to your regular Contacts database, but it includes only your favorites.

# 3
# Living on the Internet

**IN THIS PART . . .**

Surf the web from your phone.

Get to know Google's Play Store and add exciting new apps to your phone.

Chapter **7**

# You've Got the Whole (Web) World in Your Hands

I f you're like most people, one of the reasons you got a smartphone is because you want Internet access on the go. You don't want to have to wait until you get back to your laptop or desktop to find the information you need online. You want to be able to access the Internet even when you're away from a Wi-Fi hotspot — and that's exactly what you can do with your Galaxy S22 phone. In this chapter, I show you how.

The browser that comes standard with your Galaxy S22 works almost identically to the browser that's currently on your PC. You see many familiar toolbars, including the Favorites bar and search engine. And the mobile version of the browser includes tabs that allow you to open multiple Internet sessions simultaneously.

This chapter goes into much more detail on using the Internet browser on your Galaxy S22, as well as the websites you can access from your phone, and discusses some of the trade-offs you can make when viewing a web page.

# Starting the Browser

You have three options for getting access to information from the Internet via your Galaxy S22 phone. Which one you use is a personal choice. The choices are

>> **Use the regular web page.** This option involves accessing a web page via its regular address (URL) and having the page come up on your screen. The resulting text may be small.

>> **Use the mobile web page.** Almost all websites these days offer a mobile version of their regular web page. This is an abbreviated version of the full website that can be more easily read on a mobile device.

>> **Find out whether a mobile app is associated with the web page.** Many websites have found that it is most expedient to write a mobile application to access the information on its website. The app reformats the web page to fit better on a mobile screen — a convenient option if you plan to access this website regularly. I cover the trade-offs about this option later in the section "Deciding between Mobile Browsing and Mobile Apps" and explore how to find and install such apps in Chapter 8.

With this background, head to the Internet. On your Galaxy S22 phone, you may have a few choices on how to get there. Figure 7-1 shows three possible icons that can get you there. Tap any of these, and you can start surfing.

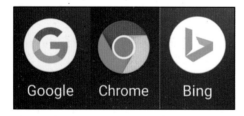

**FIGURE 7-1:** Possible paths to the Internet on your Galaxy S22.

If you want a little more understanding as to why there are multiple options, read the nearby sidebar on Internet terminology.

For your purpose, tap either the Chrome icon or the Google icon to get started. These icons will typically be on the Home screen. Alternatively, tap the Application icon and find the Chrome or Google icon.

## BASIC INTERNET TERMINOLOGY FOR DUMMIES

The term *Internet access* can mean a few different things. In some circumstances, the word Internet can mean a web browser. A *web browser* is an app that will display pages on the Internet. In other circumstances, the word Internet may refer to a search engine. You use a *search engine* to find either information you are seeking or to bring you to a website. Then you use a web browser to look at the information. Chances are the web browser on your phone that you will use is an app from Google called Chrome.

The chances are that the search engine you use on your phone is Google, but it may be the search engine for Microsoft called Bing. The goal with your phone is to get you where you want with minimal fuss. It is easier for the people who are bringing you this phone and service (Samsung, Google, and your wireless carrier) to give you three icons that essentially point to the same thing than to explain the terminology. Tapping the Internet icon, the Chrome icon, or the Google icon will most likely bring up the Chrome web browser.

**TIP**

If you love Bing, you are not out of luck. Bing is either on your phone, or you can get it installed. (I cover how to install things like Bing in Chapter 8.) For now, just stay with Chrome and Google. Things on other browsers and search engines are mostly similar, and choosing the Chrome/Google pair simplifies things.

As long as you're connected to the Internet (that is, you are either near a Wi-Fi hotspot or in an area where you have cellular service), your home page appears. If you tap the Google icon, it will be Google. If you tap Chrome, your default home page could be blank or the Google home page, but many cellular carriers set their phones' home pages to their own websites.

If you're out of coverage range or you turned off the cellular and Wi-Fi radios because you turned on Airplane mode, you get a pop-up screen letting you know that there is no Internet connection. (Read about Airplane mode in Chapter 2.)

**TIP**

If you should be in coverage but are not connected or when you get off the airplane, you can reestablish your connections by pulling down the Notification screen and either tapping the Wi-Fi icon at the top or turning off Airplane mode.

# Accessing Mobile (or Not) Websites

The browser on your phone is designed to work like the browser on your PC. At any time, you can enter a web address (URL) by tapping the text box at the top of the screen. You can try this by typing in the address of your favorite website and seeing what happens.

For example, the page shown in Figure 7-2 is the desktop version of the website Refdesk.com.

FIGURE 7-2: The desktop version of the website Refdesk.com.

As you can see, the website is all there. Also as you can see, the text is very small. This particular website is designed to take you to a lot of useful links throughout the Internet, so this is an extreme example of a regular website.

While this text is crisp and bright on your beautiful screen, it is still small. You can stretch and pinch to find the information you need. (*Stretching* and *pinching* are hand movements you can use to enlarge/shrink what you see onscreen, as covered in Chapter 2.) With a little bit of practice, you can navigate your familiar websites with ease.

The other option is to find the mobile version of a website. As a comparison, Figure 7-3 shows the mobile version of Refdesk.com. It has fewer pictures, the text is larger, and the mobile version loads faster — but it's less flashy.

**FIGURE 7-3:** The mobile version of Refdesk.com.

So how do you get to the mobile websites? If a website has a mobile version, your phone browser will usually bring it up without your having to do anything. Samsung has gone out of its way to make the web experience on the Galaxy S22 phone as familiar as possible to what you experience on your PC.

# Choosing Your Search Engine

If you tapped one of the icons shown in Figure 7-1, you opened up the browser. Now would be a good time to try some of your favorite websites. Go ahead and tap in a few website addresses and see how they look!

One of the early goals for browsers that work on smartphones was to replicate the experience of using the Internet on a PC. For the most part, the browsers on your phone achieve that goal, even if you probably need to practice your zooming and panning.

One of the key decisions for your browser is which search engine you want as the default. You probably want whatever is the search engine that exists on your desktop PC. The leading options are all really good, but you'll probably be happiest with the search engine that you use the most. If you don't care, feel free to skip to the next section on bookmarks.

1. **From the browser screen, tap the Settings icon.**

   Tapping the Settings icon, the icon that looks like a gear, brings up the menu options shown in Figure 7-4.

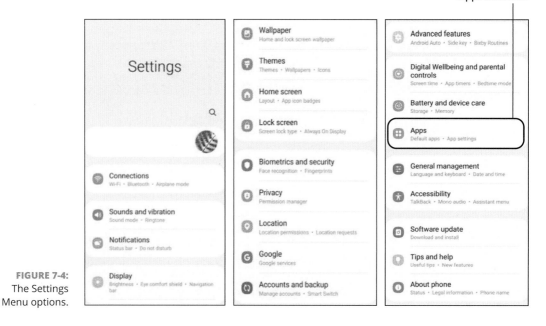

**FIGURE 7-4:**
The Settings
Menu options.

2. **Tap the Apps link.**

   Tapping the Apps link brings up the screen shown in Figure 7-5.

3. **Choose default apps link.**

   The screen shown in Figure 7-6 appears.

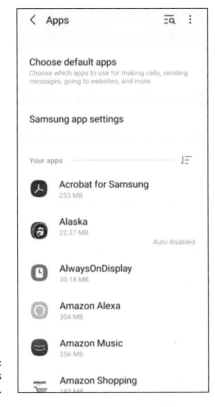

FIGURE 7-5:
The Apps
screen.

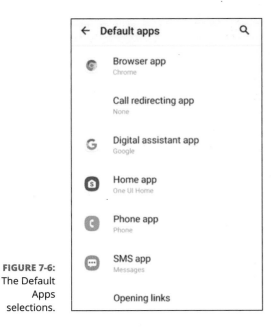

FIGURE 7-6:
The Default
Apps
selections.

4. **Tap the link for Browser App.**

   The screen shown in Figure 7-7 appears.

5. **Select the search engine you want.**

   Tapping the toggle switch by the name of the search engine you prefer takes care of it all.

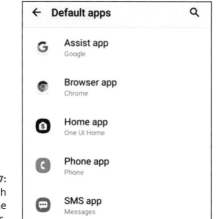

**FIGURE 7-7:** The Search Engine selections.

# Deciding between Mobile Browsing and Mobile Apps

When you open the browser, you can use any search engine you want (for example, Bing or Yahoo!). This is a very familiar approach from your experience with PCs. However, you may get very tired of all the pinching and stretching of the screen.

Many companies are aware of this. What they have done is to develop a mobile app that provides you the same information as is available on their website that is formatted for a smartphone screen. Companies in particular will put the information from their website in a mobile app and then add some important mobile features.

For example, it is very convenient to order a custom-made pizza from the website of a national pizza chain. You can order your desired variety of crusts and toppings. The website version also offers directions to the nearest store. The mobile app can take this one step further and give you turn-by-turn directions as you drive to pick up your pizza.

This capability of providing turn-by-turn directions makes no sense for a desktop PC. This capability could make sense for a laptop, but a few laptops have GPS receivers built in. It is also too cumbersome and probably unsafe to rely on turn-by-turn directions from a laptop. It is much easier to get turn-by-turn directions over a smartphone.

This is just one example where downloading a mobile app offers a superior experience. The best part is you get to choose what works for you on your Galaxy S22.

IN THIS CHAPTER

» **Getting to know Play Store**

» **Finding Play Store on your phone**

» **Seeing what Play Store has to offer**

» **Downloading and installing Facebook for Android**

» **Rating and uninstalling apps**

Chapter **8**

# Playing in Google's Play Store

One of the things that makes smartphones (such as the phones based on the Google Android platform) different from regular mobile phones is that you can download better apps than what comes standard on the phone. Most traditional cellphones come with a few simple games and basic apps. Smartphones usually come with better games and apps. For example, on your Galaxy S22 phone, you get a more sophisticated Contact Manager, an app that can play digital music (MP3s), basic maps, and texting tools.

To boot, you can download even better apps and games for phones based on the Google Android platform. Many apps are available for your Galaxy S22 phone, and that number continues to grow over time.

So where do you get all these wonderful apps? The main place to get Android apps is the Google Play Store (sometimes simply called "Google Play"). You might be happy with the apps that came with your phone, but look into the Play Store and you'll find apps you never knew you needed and suddenly won't be able to live without.

In this chapter, I introduce you to the Google Play Store and give you a taste of what you find there.

# Exploring the Play Store: The Mall for Your Phone

The Play Store is set up and run by Google, mainly for people with Android phones. Adding an app to your phone is similar to adding software to your PC. In both cases, a new app (or software) makes you more productive, adds to your convenience, and/or entertains you for hours on end — sometimes for free. Not a bad deal.

There are some important differences, however, between installing software on a PC and getting an app on a cellphone:

>> **Smartphone apps need to be more stable than computer software because of their greater potential for harm.** If you buy an app for your PC and find that it's unstable (for example, it causes your PC to crash), sure, you'll be upset. If you were to buy an unstable app for your phone, though, you could run up a huge phone bill or even take down the regional cellphone network. Can you hear me now?

>> **There are multiple smartphone configurations.** These days, it's pretty safe to assume that your computer has standard capabilities. There are differences in the amount of memory and speed of the processor. Otherwise, most PCs are similar. On the other hand, the various smartphones have significantly different features. There is a great deal of smartphone software built on the Android platform that cannot work with many Android smartphones. The Play Store ensures that the app you're buying will work with your version of phone.

# Getting to the Store

You can access the Play Store through your Galaxy S22 phone's Play Store app or through the Internet. The easiest way to access the Play Store is through the Play Store app on your Galaxy S22 phone. The icon is shown in Figure 8-1.

**TIP**

If the Play Store app isn't already on your Home screen, you can find it in your Apps list. To open it, simply tap the icon.

When you tap the Play Store icon, you're greeted by the screen looking something like what is shown in Figure 8-2.

**FIGURE 8-1:**
The Play Store
icon.

Subcategories

Top-level categories

Google Play menu button

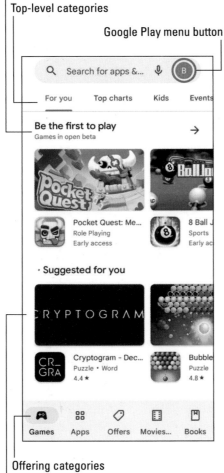

**FIGURE 8-2:**
The Play Store
home page.

Offering categories

Curated categories

As new apps become available, the highlighted apps will change, and the Home page will change from one day to the next.

In addition, the good folks at Google spend a lot of time thinking about what is the best way to help the hundreds of millions of Android users find the best application from the selection of 3.1 million apps.

This is no small task. Some of those users are very experienced and know just what they want. Others are walking in the door for the first time while still others are just coming to browse and see whether anything strikes their fancy.

The goal for Google is to make all users who comes in find what they want. The goal in this book is to give you enough information so that you can be comfortable downloading your first app and then comfortable finding other interesting apps as you become more familiar with the layout.

If working with all the categories and segments seems like too much, you can always review the 3.1 million apps alphabetically. You would start with considering the game called "AaaaaAAaaaAAAaaAAAAaAAAAA!!!" and end with the app "ZZZZZZZZZZZZ." This may take you a while. You may come to prefer one of the options I describe in this chapter.

# Seeing What's Available: Shopping for Android Apps

The panorama that exists for the Google Play home page seen in Figure 8-2 is very extensive. You can swipe to the right a long way and you can swipe down dozens of levels. Take a look at the structure of Figure 8-2.

TIP

Do not be surprised if you open up the Play Store home page one day and find that it has a completely different layout. Google tries different formats from time to time to solve one problem or another and keep things fresh. The chances are good that the lower level categories are still there, and you can find what you are looking for, even if the structure described in this chapter is no longer exactly accurate.

The first stab at helping you sort through the millions of things you can download to your S22 is the offering categories. You find these at the bottom of the page.

>> **Games:** Games are apps in which you're are an active participant. See Chapter 11 for much more information on this topic.

- » **Apps:** This is a catchall category. It includes the apps for productivity, information, social connection, or enjoyment.

- » **Offers:** Everyone likes a deal. This category brings all the sales, bundles, and sales events together in one place.

- » **Movies & TV:** The Google Play Store is a great source for video entertainment. Chapter 12 has more information on this subject.

- » **Books:** This is the section for audiobooks and e-books. Audiobooks are the smartphone version of books on tape, where a person reads you the book. E-books allow you to use your smartphone as an e-reader.

The Games option in Figure 8-2 is highlighted at the bottom of the page. If you tap one of the other categories, it will change the layout to show whatever kind of offering you're seeking.

## Navigating the Google Play apps offerings

When you've decided that you want to look at apps, you see the following options when you tap the link for one of the top-level categories:

- » **For You:** The geniuses at Google have some mathematical wizardry that tries to match you with apps that they think you would like based upon what you're already using. This is mostly a shot in the dark if you're new to the Play Store. As you use your smartphone, these guesses improve over time.

- » **Top Charts:** These are the best-selling apps. This is often a good indication that you may want to give it a try.

- » **Categories:** This option takes you to a hierarchy of app types that is useful if you already know what you want to add. There is more on these different categories later in this chapter.

- » **Kids:** If you're looking for apps suitable for kids, go here.

- » **Editors' Choice:** While relying on sales volumes in Top Charts is one way to find apps you may find valuable, this section is curated, hopefully with people who think like you.

Google is doing its best to help you find the app you need among its inventory of millions of choices. Today's version of the Play Store has a link called Categories where you can start digging by app type.

- » **Art & Design:** These apps let you exercise the powerful graphics process to make some cool images, from the abstract to the practical, such as floorplans.

- » **Augmented Reality:** These apps give you the chance to use your camera lens and see things that are not there (in a good way).

- » **Auto & Vehicles:** We love our cars. These apps help you buy them for less, enjoy them more, and enjoy them for a longer time.

- » **Beauty:** These apps offer tips and techniques on how to look that much better.

- » **Books & Reference:** These apps include a range of reference books, such as dictionaries and translation guides. Think of this section as similar to the reference section of your local library or bookstore.

- » **Business:** These apps bring you to job search sites, business that will ship your packages, and instant messages with your work cronies.

- » **Comics:** These apps are meant to be funny. Hopefully, you find something that tickles your funny bone.

- » **Education:** To quote Emil Faber, "Knowledge is good." Senator Blutarsky agrees.

- » **Entertainment:** Not games per se, but these apps are still fun: trivia, horoscopes, and frivolous noisemaking apps. (These also include Chuck Norris facts. Did you know that Chuck Norris can divide by 0?)

- » **Events:** Who doesn't enjoy the occasional blue-light special for some software you've been considering? This also brings you to sites that sell you theater tickets. New events like these are coming up every day.

- » **Finance:** This is the place to find mobile banking apps and tools to make managing your personal finances easier.

- » **Food & Drink:** Here is the place to find the best restaurants and obtain the best recipes.

- » **Health & Fitness:** This is a category for all apps related to staying healthy, including calorie counters and fitness tracking.

- » **House & Home:** Most of us like to have a place to call home. These apps help you find a home, furnish it to your liking, and keep it a comfortable temperature while minimizing your carbon footprint.

- » **Kids:** Again with the Kids. Just tell them to get off my lawn by learning something interesting on a colorful app.

- » **Libraries & Demo:** Computers of all sizes come with software libraries to take care of special functions, such as tools to manage ringtones, track app performance, and protect against malware.

- » **Lifestyle:** This category is for apps that involve recreation or special interests, like philately or bird-watching.

>> **Maps & Navigation:** Many apps tell you where you are and how to get to where you want to go. Some are updated with current conditions, and others are based on static maps that use typical travel times.

>> **Medical:** These are tools to help manage chronic conditions, such as diabetes, and buy your prescriptions at a lower cost.

>> **Music & Audio:** The Galaxy S22 comes with music and video services, but nothing says you have to like them. You may prefer offerings that are set up differently or have a selection of music that isn't available elsewhere.

>> **News & Magazines:** You'll find a variety of apps that allow you to drill down until you get just the news or weather that's more relevant to you than what's available on your extended Home screen.

>> **Parenting:** All those apps for kids must mean that there are lots of parents out there. This category is for the people who created the kids.

>> **Personalization:** These tools help you express your individuality just like everyone else.

>> **Photography:** In case the camera, photo, and video apps I cover in Chapters 9 and 10 aren't sufficient for you, here are lots more choices and add-ins.

>> **Productivity:** These apps are for money management (such as a tip calculator), voice recording (such as a stand-alone voice recorder), and time management (for example, an electronic to-do list).

>> **Shopping:** These apps give you rapid access to mobile shopping sites or allow you to do automated comparison shopping.

>> **Social:** These are the social networking sites. If you think you know them all, check here just to be sure. Of course, you'll find Facebook, LinkedIn, Twitter, and Pinterest, but you'll also find dozens of other sites that are more narrowly focused and offer apps for the convenience of their users.

>> **Sports:** You can find sports sites to tell you the latest scores and analyses in this part of the Play Store.

>> **Tools:** Some of these apps are widgets that help you with some fun capabilities. Others are more complicated and help you get more functionality from your phone.

>> **Travel & Local:** These apps are useful for traveling, including handy items, such as currency translations and travel guides. The local apps are great for staycations.

>> **Video Players & Editors:** You have the highest resolution camcorder and one of the most powerful graphics cards in your S22. These apps take the standard apps to the next level.

>> **Weather:** These are the myriad choices for finding out weather conditions in the way that suits your fancy.

Then there are the curated categories, which change over time. The Google Play Store does its best to keep these categories fresh and customized for your needs and tastes.

The Play Store's algorithms aren't always perfect. For some reason, they keep showing me curated apps related to fashion, personal hygiene, and self-grooming. This has to be a mistake.

Many of your favorite websites are now offering apps that are purpose-built for your phone. Chapter 7 talks about how you can access websites on your phone. You can use the full site with your high-resolution screen or use the mobile version. An alternative is to download the app for that website, and it will present the information you want from that website on your phone in a way that is even easier to access. In fact, when you enter a website, your phone looks to see if you have the corresponding app. If so, your phone automatically opens the app for you. Cool!

# Installing and Managing an Android App

To make the process of finding and downloading an app less abstract, I show you how to download and install one in particular as an example: the Facebook for Android app.

Follow these steps to add this app:

1. **Tap the Play Store icon, and verify that the Offering category is for Apps.**

2. **In the Query box, type** Facebook.

   Doing so brings up a drop-down screen like the one shown in Figure 8-3.

   As you can see in the search results, several options include the word Facebook. The other lines in the Apps section are for apps that include the word Facebook. These are typically for apps that enhance Facebook in their own ways — as of this moment, 112,160 of them. Rather than go through these one by one, stick with the one with the Facebook icon.

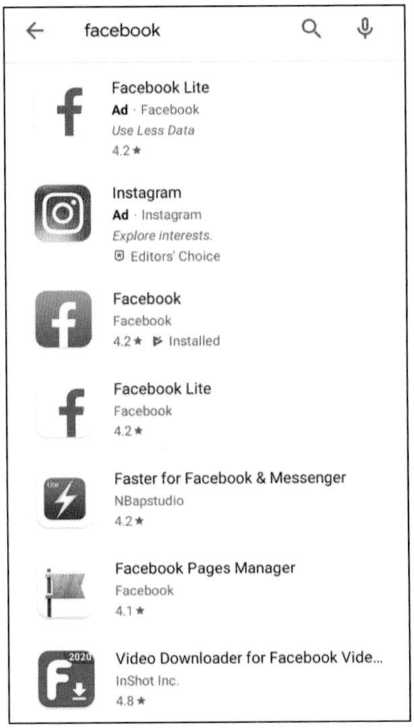

FIGURE 8-3:
The Facebook
Search results.

3. **Tap the Facebook logo.**

   When you tap the box, it brings up the screen shown in Figure 8-4.

   - **Title Line:** The top section has the formal name of the app just above the green Install button. After you click this to download and install the app, you see some other options. I give some examples later in this chapter.

   - **Description:** This tells you what the app does.

   - **Screen Captures:** These representative screens are a little too small to read, but they do add some nice color to the page.

   - **What's New:** This information is important if you have a previous version of this app. Skip this section for now.

   - **Ratings and Reviews:** This particular app has about 4.1 stars out of 5. That's not bad at all. The other numbers tell you how many folks have voted, how many have downloaded this app, the date it was released, and the size of the app in MB.

   - **Similar Apps:** Just in case you aren't sure about this particular app, the good folks at Google offer some alternatives.

- **More by Facebook:** The app developer in this case is Facebook. If you like the style of a particular developer, this section tells you what other apps that developer offers.

- **Based on Your Recent Activity:** Play Store tells you the names of other apps downloaded by the customers who downloaded this app. It's a good indicator of what else you may like.

- **Google Play Content:** This is how you tell the Play Store whether this app is naughty or nice.

Screen captures

Title

Ratings and Reviews Sections

**FIGURE 8-4:**
The Facebook app screen in panorama.

Ratings and Reviews sections

Description

4. **Tap the dark green button that says Install.**

   You see the progress of the app downloading process. When the app is all there, it begins the installation process.

   At some point in the process, most apps give you a pop-up to let you know what information from your phone that the app will use. This is to give you an idea on how this particular app may affect your privacy. An example of a permission pop-up is shown in Figure 8-5.

**Allow Facebook to access your location?**

Facebook uses this to provide more relevant and personalized experiences, like helping you to check-in, find local events and get better ads.

Deny    Allow

**FIGURE 8-5:**
The Facebook permissions screen.

Before continuing to the next step, I want to point out some important elements on this page:

Chapter 1 discusses the option to prevent an app from having access to your location information. I mention that you can allow apps to know where you are on a case-by-case basis. Here is where that issue comes up. Each app asks you for permission to access information, such as your location. If you don't want the app to use that information or share it somehow, here's where you find out whether the app uses this information. You may be able to limit the amount of location information. If you're not comfortable with that, you should decline the app in its entirety.

**TIP**

This is similar to the license agreements you sign when installing software on your PC. Hopefully, you read them all in detail and understand all the implications. In practice, you hope that it's not a problem if lots of other people have accepted these conditions. In the case of a well-known app like Facebook, you're probably safe, but you should be careful with less-popular apps.

After the app downloads and installs, you will come back to a screen like the one in Figure 8-6.

**FIGURE 8-6:**
The Play
Store app screen
for a successfully
downloaded app.

5. **Tap the dark green Open button.**

   This brings up the Home screen for Facebook, as shown in Figure 8-7.

   If you have a Facebook account already, go ahead and enter your information. Things will look very familiar. If you don't have a Facebook account, and you want to add one now, go on to the next section.

In any case, the Facebook icon appears on your Apps screen along with some other recently added apps, such as Angry Birds and Solitaire. This is seen in Figure 8-8.

If you want this app to be on your Home screen, press and hold the icon. The Facebook icon appears on your Home screen.

REMEMBER

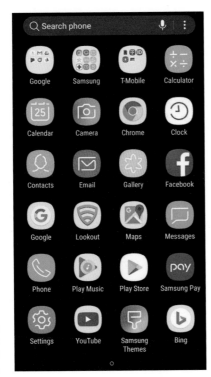

FIGURE 8-7:
The Facebook
login screen.

FIGURE 8-8:
The Facebook
icon on the
Apps screen.

# Rating or Uninstalling Your Apps

Providing feedback to an app is an important part of maintaining the strength of the Android community. It's your responsibility to rate apps honestly. (Or you can blow it off and just take advantage of the work others have put into the rating system. It's your choice.)

If you want to make your voice heard about an app, here are the steps:

1. **Open the Play Store.**

   Refer to Figure 8-2 to see the layout.

2. **Tap the Google Play Menu button (the circle with your initial).**

   Doing so brings up a drop-down menu like the one shown in Figure 8-9.

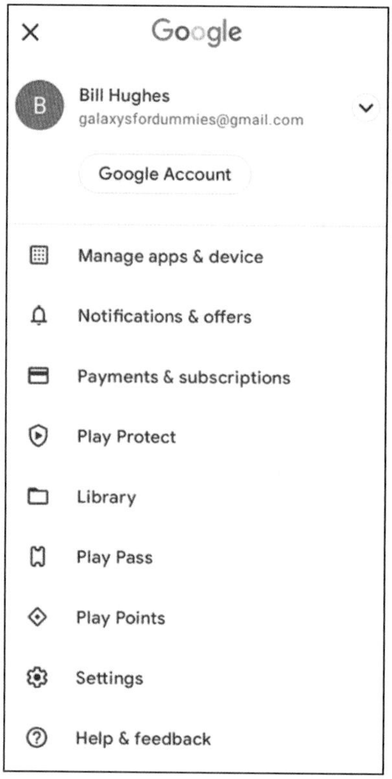

**FIGURE 8-9:** The menu from the Play Store.

**3.** **Tap the Manage Apps & Device link.**

The screen shown in Figure 8-10 appears. The Overview information is nice, but you want to manage.

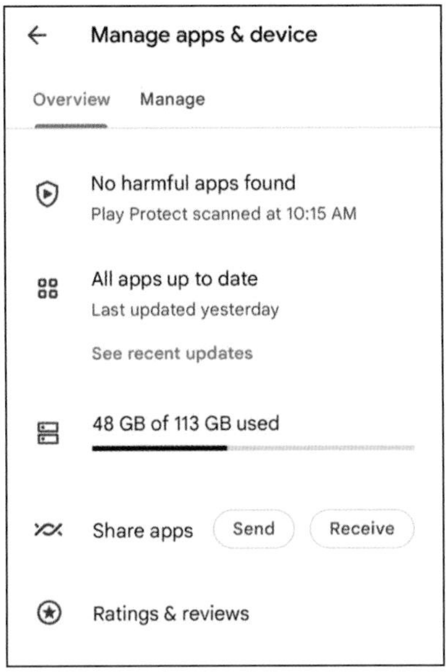

Manage apps & device

Overview    Manage

No harmful apps found
Play Protect scanned at 10:15 AM

All apps up to date
Last updated yesterday

See recent updates

48 GB of 113 GB used

Share apps    Send    Receive

Ratings & reviews

**FIGURE 8-10:** The Manage screen within the Manage Apps & Device link.

**4.** **Tap the Manage link.**

Sometimes the Manage screen opens to show the list of installed apps and the screen shown in Figure 8-11 appears, listing all the apps on your phone. Continue scrolling. You'll eventually see them all. (Other times the Manage screen opens to the Updates Available screen if you have lots of apps that are in need of updating. Other times it opens to Games.)

**5.** **Tap the Installed link.**

Tapping one of these apps is how you rate them or uninstall them, as shown in Figure 8-12.

If you love the app, rate it highly on a five star scale. To be clear, to rank it as one star, tap the leftmost star. To rank it highly with five stars, tap the rightmost star.

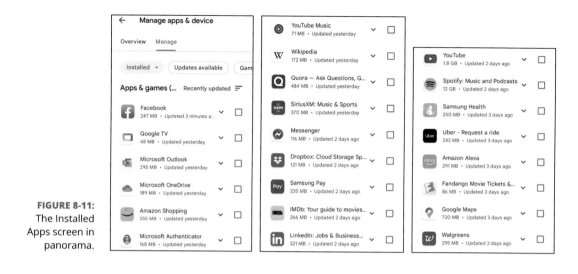

**FIGURE 8-11:**
The Installed
Apps screen in
panorama.

Uninstall button

Rating buttons

**FIGURE 8-12:**
The Apps page
for Facebook
after it is
installed.

Whatever number of stars you pick will bring up a pop-up with those number of stars, as shown in Figure 8-13. You can then tell all the world just what you think of the app.

While many app developers read the comments, it takes longer for some. I am still waiting to hear back from my comments. Please note that seeking bribes to provide positive reviews is entirely my idea, and you should probably not copy this approach. I will sue if you do.

**FIGURE 8-13:**
The Rating
pop-up.

On the other hand, if you hate the app, give it one star and blast away. It does not happen often, but there are some major league loser apps out there. In most cases, it occurs when you have an older phone, and the apps you have assume some capabilities that are not there.

This will not happen for a long, long time with the Galaxy S22, but technology does march on. Someday in the future, your wonderful phone will be a relic, and some new apps will assume that your S22 can be operated through thought control/mind reading. Alas, your S22 phone does not have that feature (as far as you know). The new app will not seem to work right.

Go ahead and give that app one star and then tap the Uninstall button. Poof. It is gone.

# 4

# Having Fun with Your Phone

# Chapter **9**

# Sharing Pictures

The Samsung Galaxy S22 should really be called a smartcamera with a phone. If you're like many mobile phone users, you love that you can shoot photographs with your phone.

You probably carry your phone with you practically everywhere you go, so you never again have to miss a great photograph because you left your camera at home.

And don't think that Samsung skimped on the camera on your Galaxy S22. In fact, they poured it on. Frankly, there has never been any smartphone like this on the market. In fact, there is so much power in the Galaxy S22 that many of the capabilities rival that of professional digital cameras. There are three cameras on the back: The first is a 50-megapixel camera with wide-angle lens, the second is a 12-megapixel camera with an ultra-wide-angle lens, and the third is a 10-megapixel camera with a telephoto lens. With the Galaxy S22, you can zoom in by a factor of 30x to make an object that is 100 yards away appear as if you're 10 feet away.

In addition, there is a 10-megapixel camera on the front of the phone. This is sometimes called the "selfie lens," because it's easy to position the camera while looking at the screen. Don't turn up your nose at a mere 10-megapixel camera — it's more than enough for most selfies.

Don't get me started on the capabilities of the S22 Ultra! A 12-megapixel camera is adequate for most people, but the camera with the wide-angle lens in the S22

Ultra boasts 108 megapixels. There are other awe-inspiring camera capabilities in the S22 Ultra. For example, you can zoom in by a factor of 100x and have the ability to take shots in extremely low light.

I could toss around specifications that would impress any scientist or professional photographer. I could explain the calculations needed to pick the better choice about when to use the wide-angle and the ultra-wide-angle lenses to get the optimal field-of-view.

The good news is that the good folks at Samsung get it and have taken steps to take all this raw power and make it easy. And if you want to get fancy, you have that option, although it may take a moment or two.

After you've taken your images, you can view them on that wicked Super AMO-LED screen. Samsung also includes a Gallery app for organizing and sharing. As a practical matter, it makes sense to view most of photos on your phone. For most of us, that's more convenient than making prints and putting them in a scrapbook. Your phone is always with you, and if someone wants a copy, you just tap a few icons and it's in that person's photo album.

It's truly amazing to consider the number of options that you have for your photographs. Samsung and a bevy of third-party software developers keep adding new options, filters, and ways to share these files. It would not be surprising if your phone now has more capabilities than your digital camera.

These capabilities actually cause a problem. There are so many options that it can be overwhelming. Research has shown that people fall into one of three categories: The first group tends to use the default settings. If they want to alter the image, they'd prefer to do it on their own PCs. The second group likes to explore some of the capabilities on the phone and will have some fun looking at the scenarios, but keep things within reason. The third group goes nuts with all the capabilities of the phone.

To accommodate everyone, this chapter starts with the basics. I cover how to use the camera on your phone, view your pictures, and share them online. This works for everybody.

I then focus on the most popular settings for the second group. There are some important capabilities. With all these exotic lenses, there is some awesome intelligence embedded in the Camera app. Without your needing to do anything, it takes its cues from what is on the screen and will pick the best lens for what it thinks you're trying to do.

For the third group, the best that can be done is to lead you in the right direction. As much as I would like to, it would be impossible to cover all the options and combinations. In my count, there are 2.43 billion possible combinations of filters, lighting, and modes. If you were to start now and take a picture every 10 seconds, you would not run out of combinations for over 700 years. This would not even include all the options for sharing these images.

Realistically, neither you nor your phone will last that long. To keep things in the realm of reality, I introduce only the most important and valuable options and go from there.

# Say Cheese! Taking a Picture with Your Phone

Before you can take a picture, you have to open the Camera app. The traditional way is to simply access the Camera app from the Application list. Just tap the Camera icon to launch the app.

 Because the camera is so important, here are a few more ways to get to the Camera app. First, press the Power button twice. Boom. There it is. (If this happens to be turned off on your phone, you can toggle the Quick Launch Camera in the Advanced Features within Settings.)

Next, unless you have turned off the capability for security purposes, there is a camera icon on your lock screen. If you swipe the icon across the screen, the Camera app bypasses the security setting. You can snap away, but you can't access the photo gallery or any other files.

 Here is a suggestion. If you see something suspicious, but are not ready to call 911, go ahead and start taking photos of what concerns you. If you need to, you can go back to the lock screen and slide the phone icon to the right to call 911, again without needing to unlock.

TIP

A closely related app on your phone is the Gallery, which is where your phone stores your images. The icons for these two apps are shown in the margin.

 With the Camera app open, you're ready to take a picture within a second or two. The screen becomes your viewfinder. You see a screen like the one shown in Figure 9-1.

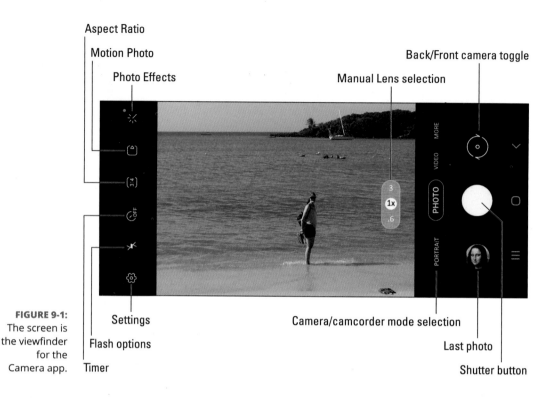

Aspect Ratio

Motion Photo

Photo Effects

Back/Front camera toggle

Manual Lens selection

**FIGURE 9-1:**
The screen is
the viewfinder
for the
Camera app.

Settings

Flash options

Timer

Camera/camcorder mode selection

Last photo

Shutter button

# A SUPER-FAST PRIMER ON SUPER AMOLED

All Samsung Galaxy S22 phones have a Super AMOLED screen. Allow me to take a moment to explain what makes this so good — and what makes you so smart for having bought the Samsung Galaxy S22.

To start, think about a typical LCD screen, like what your TV or PC might have. LCDs are great, but they work best indoors where it's not too bright. LCDs need a backlight (fluorescent, most commonly), and the backlight draws a fair amount of power, although much less power than a CRT. When used on a phone, an LCD screen is the largest single user of battery life, using more power than the radios or the processor. Also, because of the backlight on an LCD screen, these screens display black as kinda washed-out, not a true black.

The next step has been to use light-emitting diodes (LEDs), which convert energy to light more efficiently. Monochrome LEDs have been used for decades. They are also used in the mongo-screens (jumbotrons) in sports arenas. Until recently, getting the colors right was a struggle. (Blue was a big problem.) That problem was solved by using organic materials (*organic* as in carbon-based, as opposed to being grown with no pesticides) for LEDs.

The first organic LEDs (OLEDs) looked good, drew less power, and offered really dark blacks, but still had two problems. Their imaging really stank in bright light, even worse than did LCD screens. Also, there was a problem with *crosstalk* — individual pixels would get confused, over time about whether they were on or off. You'd see green or red pixels remaining onscreen, even if the area was clearly supposed to be dark. It was very distracting.

The solution to the pixels' confusion is called *Active Matrix,* which tells the pixels more frequently whether they are to be on or off. When you have Active Matrix technology, you have an Active Matrix Organic LED, or AMOLED. The image you get with this technology still stinks in bright light, but at least it's some improvement.

Enter the Super AMOLED technology, made by Samsung. When compared with the first AMOLED screens, Super AMOLED screens are 20 percent brighter, use 20 percent less power, and cut sunlight reflection by 80 percent. This is really super!

And how do you snap the picture? Just tap the big white button on the right, which is the digital shutter button. The image in your viewfinder turns into a digital image.

This screen still uses a significant share of battery life, but less than with earlier technologies. With Super AMOLED, you even save more power if you use darker backgrounds where possible. A few picture–taking options are right there on the viewfinder. Going from the viewfinder clockwise around the screen, the options include

>> **The viewfinder:** The viewfinder shows what will be in the picture. Okay, this is obvious, but this is also one way you control the magnification. From within the viewfinder, you can zoom in and out by pinching or stretching the screen.

- Stretch the screen to zoom in.

- Pinch the screen to zoom out.

>> **Lens Selection:** You can zoom in by stretching or zoom out by pinching. Or you can tap one of the lens options. In Figure 9-1, the 3 is a telephoto with 3x magnification, the circle with the 1x is normal with 1x magnification, and the .6 is for the macro lens. Here's where there is a difference among the Galaxy S22 models. There are a different number of front and back lenses based upon the model you have and whether you're using the back- or front-facing camera. You'll see a different number of options based upon the lenses you have. This is where the extra cost pays off if you have an S22 Ultra.

» **Rear-facing camera to front-facing camera toggle:** This icon takes you from the rear-facing camera to the not-too-shabby 10-megapixel front-facing camera. As mentioned, this is a good option for selfies because you can see what the shot will look like before you snap it. If you have the S22 Ultra, the front-facing camera is a staggering 40-megapixel monster.

» **Shutter button:** This icon takes the picture.

» **Gallery:** Tap this icon to see the pictures you've just taken. I discuss the Gallery in the section "Managing Your Photo Images," later in this chapter.

» **Camera/Camcorder mode:** These are important. Figure 9-1 is set for Photo mode, which is probably the best place to start. Your camera is super-smart, and Photo mode does a good job of guessing how to make your image look its best. However, if you want to get a little fancy — or really fancy — you can change the settings. You can find more on this option later in this chapter. This is also how you switch to recording video. More on this later in Chapter 10.

» **Settings:** Settings also gives some fancy options that are covered in the next section.

» **Flash options:** Even if you want to keep it simple, you'll want to know how to control your flash. Sometimes you need a flash for your photo. Sometimes you need a flash for your photo, but it's not allowed, as when you're taking images of fish in an aquarium, newborns, or some animals. (Remember what happened to King Kong?) Regardless of the situation, your phone gives you control of the flash. Tap the Settings icon at the top of the viewfinder. The options are Off, On, or Auto Flash, which lets the light meter within the camera decide whether a flash is necessary. The options are shown in Figure 9-2.

» **Timer:** This creates a few seconds lag between the time you press the shutter button and when it takes the image so you have time to get into a group picture.

» **Aspect Ratio:** The standard photo is typically 4 units wide by 3 units high. This means that it has a 4:3 aspect ratio. This is conventional, but why be conventional all the time? You can play with different options.

Flash always off          Flash always on

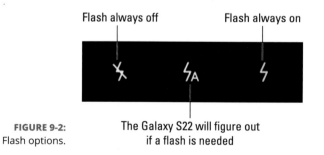

**FIGURE 9-2:**
Flash options.

The Galaxy S22 will figure out
if a flash is needed

>> **Motion Photo:** I cover this in more detail in the "Using the Photo mode settings" section, later in this chapter.

>> **Photo Effects:** I cover this in more detail in the "Using the Photo mode settings" section, later in this chapter.

Also mentioned earlier is the ability to zoom. You accomplish this by using the stretch or pinch motion on the screen. To zoom in, you start with the normal image as shown in the left image of Figure 9-3. Then you put your fingers any-where on the screen and stretch. That zooms you in, as shown in the right image of Figure 9-3.

**FIGURE 9-3:**
The viewfinder
when zooming.

The screen shows you the level of zoom and markings to show you how far you can go. There are also shortcuts underneath the image if you want to just jump to higher magnification. This screen show the maximum of 30x magnification. This image shows a modest 1.9x magnification.

When you jump to 30x, you can see the person admiring the scenery on the observation deck of the Space Needle. If you have the S22 Ultra and take it up to 100x, you can see that he has walked 3,296 steps today on his Fitbit (okay, not really, but 100x is still pretty amazing).

# FOR SKEPTICS ONLY

If you've ever used a cameraphone, you might be thinking, "Why make such a big deal about this phone's camera? Cameraphones aren't worth the megapixels they're made of." Granted, in the past, many cameraphones weren't quite as good as a digital camera, but Samsung has addressed these issues with the Galaxy S22. You may as well try to sell your digital camera at a garage sale because your S22 takes much better shots:

- **Resolution:** The resolution on most cameraphones was lower than what you typically get on a dedicated digital camera. The Galaxy S22, however, sports at least a 12-megapixel camera — and that's good enough to produce an 8 x 10-inch print that's indistinguishable from what you could produce with a film camera, and 4 x 6-inch photos are plenty large for most uses.

- **Low light:** Many cameraphones perform badly in low light conditions. The S22 works incredibly well in low light.

- **AutoFocus and artificial intelligence:** We amateurs like it when a camera will do the hard work of focusing. The technology in the S22 takes a big leap by having more pixels figuring out where to focus, and then there are the smarts to figure out how to get the most out of what you see in the viewfinder.

- **Photo transfer:** With most cameraphones, the photos are hard to move from the camera to a computer. With the Samsung Galaxy S22, however (remember: it uses the Android operating system), you can quickly send an image, or a bunch of images, anywhere you want, easily and wirelessly.

- **Screen resolution:** In practice, many cameraphone users just end up showing their pictures to friends right on their phones. Many cameraphone screens, however, don't have very good resolution, which means your images don't look so hot when you want to show them off to your friends. The good news is that the Samsung Galaxy S22 has a bright, high-resolution screen. Photos look really good on the Super AMOLED screen.

- **Organization:** Most cameraphones don't offer much in the way of organizational tools. Your images are all just there on your phone, without any structure. But the Samsung Galaxy S22 has the Gallery app that makes organizing your photos easier. It's also set up to share these photos easily.

You probably know that it's not a good idea to touch the lens of a camera. At the same time, it's practically impossible to avoid touching the lenses on your Galaxy S22. This can create a problem where there can be a buildup of oil and dirt on your high-resolution lens. You should clean the lens of your camera from time to time with a microfiber cloth to get the most out of your camera. Otherwise, your high-resolution images might look like you live in a perpetual fog bank.

After you take a picture, you have a choice. The image is automatically stored in another app: the Gallery. This allows you to keep on snapping away and come back to the Gallery when you have time. I cover the Gallery app more in the upcoming section "Managing Your Photo Images."

However, if you want to send that image right away, here's what you do:

1.  **From the viewfinder screen, tap the Gallery icon.**

    The viewfinder shows the Gallery icon at the top-right corner of the viewfinder. When you tap it, it brings up the Gallery app, as shown in Figure 9-4.

    This brings up the current image, along with the some other recent photos.

**FIGURE 9-4:** The Gallery app.

## 2. Tap the thumbnail of the image you want to share.

It also brings up some options, as seen in Figure 9-5. Right now, you're interested in the Share option.

**FIGURE 9-5:** Gallery options for the current image.

Share Icon

## 3. Tap the Share button.

This brings up the options you can use to forward the image; see Figure 9-6 (although your phone might not support all the options listed here and may have a few others not in this image). You have the following options (not necessarily in order):

- **Nearby Share:** You can send the picture to another smartphone that also has Nearby Share enabled. (Having only friends who have a recent Samsung Galaxy phone is your best bet to be able to use this option.)

- **People you recently texted:** You can send the picture to the last four people with whom you texted.

- **Email:** You can send the image as an attachment with your primary email account.

- **Facebook:** You can take a picture and post it on your Facebook account with this option.

- **Messaging:** You can send the picture immediately to someone's phone as a text message.

- **Bluetooth:** You can send images to devices, such as a laptop or phone, linked with a Bluetooth connection.

- **Gmail:** If your main email is with Gmail, this option and the Email option are the same.

- **Your cloud provider:** Figure 9-6 shows Google Drive, but if you sign up for any of the cloud storage options, they'll be here.

- **Wi-Fi Direct:** Talk about slick! This option turns your phone into a Wi-Fi access point so that your Wi-Fi–enabled PC or another smartphone can establish a connection with you.

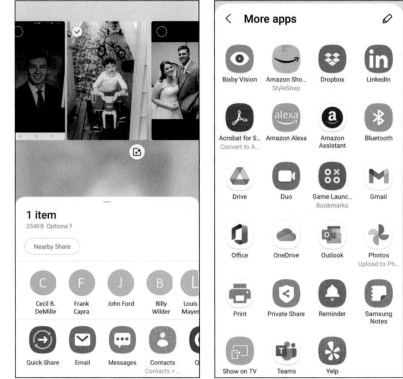

**FIGURE 9-6:**
Sharing options for the current image.

The point is that there is an overabundance of options. If an app on your device works with images, this is the place you can upload that image. If one of these options doesn't quite suit your need to share your pictures, perhaps you're being too picky!

# Getting a Little Fancier with Your Camera

Using the default Camera setting, Photo, to snap pics is perfectly fine for those candid, casual, on-the-go shots — say, friends in your living room. However, your Samsung Galaxy S22 phone camera can support much more sophisticated shots. Your digital SLR camera may have a bigger lens than your phone, but I can assure you that your phone has a much bigger brain than your camera. I suggest that even the users who want these default settings get familiar with the camera's Mode settings. These are easy to access and will improve the image quality with minimal effort.

## Using the Photo mode settings

The mode setting is where you make some basic settings that describe the situation under which you will be taking your shot. The default is a single picture in Photo mode, as shown in Figure 9-1. Sliding this icon to the left brings up the options associated with the MORE link, shown in Figure 9-7.

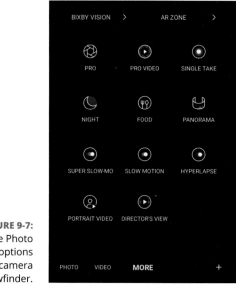

**FIGURE 9-7:**
The Photo mode options on the camera viewfinder.

The options on the top of the viewfinder bring up a number of choices for photos:

>> **Pro:** This mode gets you to all kinds of camera settings that someone who knows what they're doing would want to control rather than let the camera make the choice. This includes setting the exposure, shutter speed, ISO sensitivity, white balance, focal length, and color tone. If you don't know what these terms mean, this is not the mode for you.

>> **Single Take:** This feature is so cool, it's almost cheating. When you enable this option and tap the shutter button once, the camera uses all its lenses and modes using all the styles and picks the best one. How does it know the best one? Well, the S22 is so smart it can figure it out.

>> **Night:** This mode takes pictures when it's dark, and I mean darker than you ever thought possible.

>> **Food:** This mode takes pictures that emphasize the colors in food to make your friends even more jealous (and/or hungry).

>> **Panorama:** This mode lets you take a wider shot than you can with a single shot. Press the Camera button while you rotate through your desired field of view. The app then digitally stitches the individual photos into a single wide-angle shot.

Chapter 10 covers the options for Pro Video, Super Slow-mo, Slow Motion, Hyper-lapse, Portrait Video, and Director's View. Otherwise, choose the option that sounds right by flicking through the options, and snap away.

## Settings options on the viewfinder

Tapping the Settings icon on the viewfinder brings up a number of choices, shown in Figure 9-8:

>> **Scene Optimizer:** Once again, I will point out that your phone is super-smart. This setting automatically adjusts setting to take the best possible picture. You can turn off this option, but why would you want to?

>> **Shot Suggestions:** This takes the scene optimizer one step further. Your phone will suggest how to set up the shot, and when it's perfect, it will take the picture for you. Try it and see if you like this one.

>> **Scan QR Codes:** When the viewfinder sees a QR code, it will try to figure out what it says. If you don't care, switch off the toggle button.

» **Swipe Shutter Button To:** You have two options here. The default when you swipe the shutter button to the side is to take a quick series of pictures. That way, you can go back and pick the best one to keep. You could also set it so that you just take a single picture and store the image as a GIF. If you don't know what a GIF is, you probably don't want to bother with this option.

» **Picture formats:** Your average smartphone will save your photo in one of the popular photo formats. But this isn't your average smartphone. You can change the default option to High Efficiency (HEIF) to save some storage. You can select RAW copies if you have super-duper photo-editing software back on your PC. You may want to leave this one alone, though — it's complicated.

» **Save Selfies as Previewed:** If you've ever taken a selfie, you may have noticed that it was a mirror image of what you expected. If that doesn't bother you, toggle this option on. Likewise, if the wording on your shirt says "WOT TOW" or "I bid I" you're all set. Otherwise, let you phone switch your selfie so the words aren't reversed.

» **Selfie Color Tone:** You can go natural, or you can have a bright, shiny face.

» **Auto HDR:** The camera can apply this image enhancement when it's needed. You can also just leave it on all the time. Your choice.

» **Tracking Auto-Focus:** The Samsung Galaxy S22 can spot a person's face and assume that you want it to be the place where you focus if you toggle this option to the On position. Otherwise, if you don't use this mode, the camera may assume that you want whatever is in the center of the viewfinder to be in focus.

» **Grid Lines:** Some people like to have a 3 x 3 grid on the viewfinder to help frame the shot. If you are one of these people, toggle on this option.

» **Location Tags:** The Samsung Galaxy S22 uses its GPS to tell the location of where you took the shot. If this is too intrusive, you can leave out this information on the image description.

» **Shooting Methods:** This menu gives you some convenience options. For example, you can choose what pressing on the volume buttons does: Take a picture or video, zoom in or out, or control the volume. You can also have the phone take a picture with voice commands, relocate the shutter button to somewhere else on the screen, or take the picture when you show the camera the palm of your hand.

» **Settings to Keep:** It can be annoying when you select the camera options you want one day, and then later, you need to remember what you did. Problem solved with this option. You tell the camera to keep your settings and off you go.

>> **Shutter Sound:** It can be satisfying to hear the click when you take a picture. If you would rather not, turn that sound off here.

>> **Vibration Feedback:** In addition to the click, you can have that feeling you remember back in the day when your analog single-lens reflex (SLR) camera would shake when you pressed the button. Your phone can't bring back Pet Rocks or dot matrix printers, but it can bring back that visceral feeling of knowing that you got the shot.

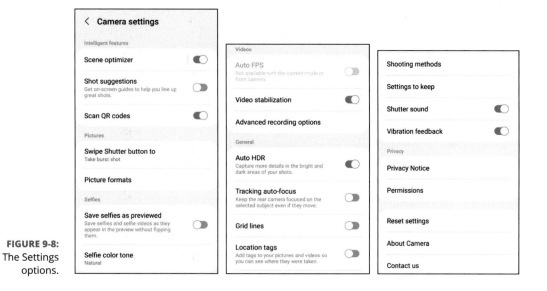

**FIGURE 9-8:**
The Settings options.

## Photo Effects options

When you tap the magic wand on the viewfinder, you get to see a number of effects that you can apply to your image. Some examples are shown in Figure 9-9.

Select the option that looks right for your creative vision by tapping on the thumbnail and snap away.

**FIGURE 9-9:**
Some examples of Effects options.

# Visualizing What You Can Do with Bixby Vision

In Figure 9-7, you may have noticed an option called Bixby Vision in the viewfinder in the upper-left corner. Bixby is Samsung's version of Apple's Siri or Google's Alexa. This effort ultimately didn't pan out as well as those other attempts, but that doesn't mean that it isn't as cool as its more successful cousins.

Bixby Vision can tell you information about what the camera is pointed at:

>> If it's an object, Bixby Vision can find similar images on Pinterest.

>> If it's text, Bixby Vision can read it so you can copy it into an app.

>> If it's text in a foreign language, Bixby Vision can read it and convert it to text. You can then translate that text into any of 108 languages.

>> If it's a wine label, Bixby Vision can tell you the ranking of the wine.

How Bixby Vision works is simple. You open the camera application, click the MORE option, and tap Bixby Vision. You then get some options that let you tell Bixby what you want to do (see Figure 9-10).

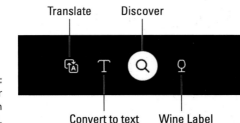

**FIGURE 9-10:**
The options for
Bixby Vision in
the Viewfinder.

Translate    Discover

Convert to text    Wine Label

Let's say you want to translate some foreign text into English. You slide the selector over to the *T* for Translator and center the viewfinder over the text. Figure 9-11 shows the original text and then the translation.

### Original in French

C'était LE MEILLEUR des temps, c'était le pire des temps, c'était l'âge de la sagesse, c'était l'âge de la folie, c'était l'époque de la croyance, c'était l'époque de l'incrédulité, c'était la saison de la Lumière, c'était la saison des Ténèbres, c'était le printemps de l'espérance, c'était l'hiver du désespoir, nous avions tout devant nous , nous n'avions rien devant nous, nous allions tous directement au Ciel, nous allions tous directement dans l'autre sens — bref, la période était si loin comme la période actuelle, que certaines de ses autorités les plus bruyantes insistaient pour qu'elle soit reçue, pour le bien ou pour le mal, dans le degré de comparaison superlatif seulement.

### Bixby Translation to English

**FIGURE 9-11:**
The Bixby
Vision
translation of a
familiar novel
from your high
school days.

It was THE BEST of times, it was the worst of times, it was the age of wisdom, it wa s the age of madness, it was the age of belief, it was the age of disbelief, it was t he season of Light, it was the season of Darkness, it was the spring of despair, had everything before us, we had nothing in front of us, we were all going direct to Heaven, we in short, the period was so distant as the present period, that we could all go directly the other way, some of the loudest authorities insisted that it be received, for the good or for the sake of bad, in the degree of superlative comparison only.

It isn't perfect, but it's not too bad. It'll help you out in a pinch and do a pretty good job of it. Let's see how well Alexa does in translating some text!

# Managing Your Photo Images

After you take some great pictures, you need to figure out what to do with them. Although you can send an image immediately to another site or via email, it will likely be the exception.

In most cases, it's easier to keep on doing what you were doing and go back to the Gallery app when you have some time to look at the images and then decide what to do with them. Your choices include

>> Store them on your phone within the Gallery app.

>> Transfer them to your PC to your photo album application by sending them with email.

>> Store them on an Internet site, like Google Photos or Flickr.

>> Print them from your PC.

>> Email or text them to your friends and family.

>> Any combination of the preceding choices.

Unlike many regular phones with a built-in camera, the Galaxy S22 makes it easy to access these choices. You need to determine the approach you want to take to keep your images when you want them to stick around. The rest of this chapter goes through your options.

**REMEMBER**

Even though the Camera app and the Gallery app are closely related, they are two separate apps. Be sure that you keep straight which app you want to use.

The Gallery Home screen (refer to Figure 9-4) shows how the app first sorts the images on your phone into folders, depending upon when they originated.

All your photos from the Camera app are placed in files sorted by date. The app takes a shot at grouping them when a series of pictures or videos are taken about the same time.

# Using Images on Your Phone

In addition to sharing photos from your camera, your Galaxy S22 phone allows you to use a Gallery photo as wallpaper or as a photo for a contact. And if the fancy shooting settings in the Camera app aren't enough, you can wrangle minor edits — as in cropping or rotating — when you have an image in the Gallery app.

The first step is to find the image that you want in Gallery. If you want to do something to this image other than send it, tap the Edit button (a silhouette of a pencil) at the bottom of the screen (refer to Figure 9-5). Some of the options include

- **Slideshow:** This displays each of the images for a few seconds. You can not only set the speed for transition, but also add music and select among several image transitions.

- **Crop:** Cut away unnecessary or distracting parts of the image. The app creates a virtual box around what it considers to be the main object. You can move the box around the image, but you cannot resize it. You then can either save this cropped image or discard.

- **Set As:** Make this image your wallpaper or set it as the image for a contact.

- **Print:** This option allows you to print if you have set up a local printer to communicate with your phone either through Wi-Fi or Bluetooth.

- **Rename:** These options allow you to rotate the image right or left. This is useful if you turned the camera sideways to get more height in your shot and now want to turn the image to landscape (or vice versa).

- **Details:** See the information on the image — its metadata, which is fixed and cannot change.

# Deleting Images on Your Phone

Not all the images on your phone are keepers. In fact, you may get accustomed to deleting far more pictures than you keep. This is hard for some of us who are used to expensive film. However, before too long, you'll have far too many pictures, which defeats the purpose.

When you want to get rid of an image, press and hold the image you want to delete. In a second, a check box with the image selected will appear. Also, the links appear at the top to either Share or Delete. If you want to delete this image, tap Delete. The camera verifies that this is your intent. After you confirm, the image goes away.

If you want to delete more images, you can tap all the images you want to make go away. It is selected if it has a green checkmark on the image. Tap away, then hit Delete. It will confirm with you once. Tap again and these images are gone forever.

**WARNING**

When I say that the photos you delete are gone forever, I do mean *for-ev-er*. Most of us have inadvertently deleted the only copy of an image from a PC or a digital camera. That's not a pleasant feeling, so be careful.

Chapter **10**

# Creating Videos

The Samsung Galaxy S22 really poured it on with the still images. Chapter 9 covers all the amazing capabilities that your phone can do with photographs. It's truly amazing. But wait! There's more — much, much more! You can take amazing videos with your phone.

However, before I jump into the capabilities, take a moment and think about what you want to do with videos. A knee-jerk reaction is: "I want the highest resolution and the most frames per second available. After all, memory is cheap."

The issue to consider is that recording videos uses lots of memory and other resources. Plus, videos don't lend themselves to easy modification after you take them. For example, you can take a photo on your phone. You can then make changes, such as adjusting the brightness and cropping it, after the fact.

Making similar adjustments to videos is impractical for most of us. If you take a video in super slo-mo, you can't convert it to normal speed without very costly video-editing software. The point is that video images are fun, but you need to think through in advance what you want to achieve.

You may just want to capture the fun at the beach or a party with your friends and family. That's fairly typical. It can be more fun than still photography. You can also use your phone as a dash-cam with the hope of capturing an accident. That's also fairly typical.

This first part of this chapter presents the basic of taking videos like these. I cover how you set it up, some of the basic options, and how you can view your wonderful creation. Later in the chapter, I discuss some of the more exotic capabilities of the S22.

# Ready . . . Action! Taking a Video with Your Phone

To take a video, you start with the Camera app.

With the Camera app open, you're ready to take a picture. But you want to take a video. To start the video, you swipe the Photo link to the left (which is down in the landscape orientation), shown in Figure 10-1.

Image stabilizer

Video Format

Auto framing on

Record button

**FIGURE 10-1:**
The screen is the viewfinder for the Camera app in Video mode.

This is very similar to the screen in Figure 9-1. There are four differences for the icons on the viewfinder:

>> **Record Button:** Instead of a shutter button, there is the Record button.

>> **Video Stabilizer:** This option does some digital magic and makes it look like you're moving the phone smoothly from place to place. If you shut off this capability, the video will look jerky and will be hard to watch. Under normal circumstances, it's best to leave the Video Stabilizer on. Your viewing audience will appreciate it.

>> **Auto Framing:** This option automatically zooms in or pans out to include all the people it sees in the frame. It saves you the social risk of cutting one of your friends in half. It could also drive you nuts if you want control of what's on the screen.

>> **Video Format:** This option gives you quick access to numerous options for image resolution and frames per second.

All you need to do to get started is push the Record button. The image in your viewfinder turns into a video, and you see the icon switch from the image on the left in Figure 10-2 to the icon on the right.

Record
button

Pause or stop
button

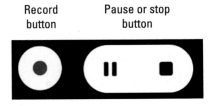

**FIGURE 10-2:**
The Record button, Pause button, and Stop button in the viewfinder.

Pausing allow you to temporarily stop recording the video and then restart. Stopping saves the video as it is. It saves the files and reverts to the Record button. If you want to record again, it creates a new file, which is often no big deal.

To get to the video, from the viewfinder screen, you simply tap the Gallery icon. The viewfinder shows the Gallery icon next to the Record button. When you tap it, it brings up the Gallery app, as shown in Figure 10-3.

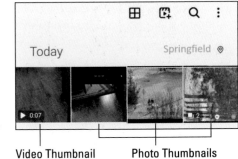

**FIGURE 10-3:**
The Gallery app with a video.

Video Thumbnail          Photo Thumbnails

This brings up the video along with the other recent photos. You can tell the videos from the photos because the videos have the play icon, which is an arrowhead pointing to the right, along with information on the duration of the video. In this case, the video is 7 seconds.

You have a few choices at this point. If you want to view the video, tap the thumbnail. It will expand the thumbnail to the full screen. Tap the link that says Play Video, and off it goes. Just like viewing photos, watching your video on your phone is a pleasant experience.

At the same time, a really good video is really, really worth sharing. To share your video, you press and hold the thumbnail. Every thumbnail suddenly has a white circle in the upper-left corner. Tap the circle with the file you want and a check mark appears (see Figure 10-4).

**FIGURE 10-4:**
Selecting the
current video.

You now have two options: You can share it or you can delete it. Since this is really, really worth sharing, you tap the share option. This brings up the sharing screen (see Figure 10-5).

Yes, this is the same screen used as sharing photos, but bear with me. Good videos should be shared on social media without delay. Go ahead and tap the Facebook icon that has the title Your Story. That brings a screen up like seen in Figure 10-6.

When you tap the Share button, it will verify a few things, such as validating your desired level of privacy. Accept it all and this video of a cute kid riding on his trike (or whatever is in your video) will be posted for the world to see.

**FIGURE 10-5:**
Sharing
options for the
current video.

**FIGURE 10-6:**
The Facebook
Share to Story
option.

# Taking Videography to the Next Level

The default video settings do the job in the vast majority of cases. But that is boring. You have a number of exotic options that you may as well take out for a test drive and see how it goes.

## The Video mode settings

Do you remember the part of the viewfinder where you moved from Photo mode to Video mode? If you keep going, you hit the More option. Sliding to this icon brings up the options shown in Figure 10-7.

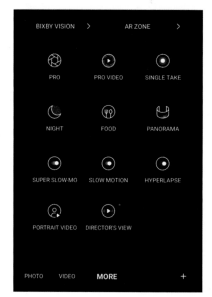

**FIGURE 10-7:**
More Video options on the camera viewfinder.

The options in the viewfinder bring up a number of choices:

» **Pro Video:** This option opens up all kinds of video options.

» **Super Slow-Mo:** This switches you to video mode to slow down really fast shots. The exceptional thing is that your S22 does this in full 4K resolution, so you're getting both high speed and high resolution with no compromise (other than filling up your memory card)!

» **Slow Motion:** This is plain-old, boring slow motion. It's still cool.

- » **Hyperlapse:** You can use this setting to create a time-lapse video. The video camera will adjust the speed of the shots based upon the movement of the phone.

- » **Portrait Video:** This setting lets you get very close, within a few feet, of an object. The normal autofocus does not handle it well. If you hear the words, "I'm ready for my closeup, Mr. DeMille," you can be assured that your shot of Norma Desmond, played by Gloria Swanson, will look great *and* be in focus.

- » **Director's View:** This is a cool option. If you want to be the envy of Cecil B. DeMille, try this option. In addition to the viewfinder view of your video, the other lenses show up as thumbnails. This gives you the creative ability to jump from the wide-angle lens to the telephoto lens while you're shooting.

Go ahead and experiment with these, so when the opportunity presents itself, you know what option to choose.

## Settings options on the viewfinder

The Settings icon on the viewfinder brings up choices for the video. Before I get into these, I cover some important considerations.

The first choice you need to make is the aspect ratio. All the cool kids have displays that are 16:9. If you're of a certain age, you remember standard televisions, which were 4:3. Forget about that old technology. Your videos should be 16:9, unless you're getting carried away and are getting artsy.

The next consideration is what resolution you want. The most you can select is 8K. Only the coolest of the cool kids have 8K screens these days. These are expensive and there is very little content for it . . . yet. Even the mainstream 4K TVs can have images that are a little too sharp for some people. However, 4K TVs have a lot of content these days, so they're selling lots of TVs like this.

Still, the variations of HD are pretty darn good. The starting point on the variations of HD are Standard HD, which is 1280 x 720 pixels. This is sometimes referred to just as 720 HD. Then there is Full HD, which is 1920 x 1080 pixels on a screen. This is sometimes called 1080 HD. Between the 4K TVs and the 1080 HD is Ultra HD (UHD). You access these options (shown in Figure 10-8) by tapping the Video Format icon shown in Figure 10-1.

**FIGURE 10-8:**
The Video Format options.

So, which option is the best? Usually more resolution is good, but at some point, it just takes up a lot more memory. Try them all and see what you like. My guess is that Full HD is probably fine.

Then, the next consideration are the frames per second. You have the option of standard, or you can choose 60 frames per second (fps). But why? Even if you bought the S22 with the extra memory, this option eats it up like crazy. Go ahead and ignore that setting, and you won't regret it . . . probably.

With that background, the video options within the camera settings are shown in Figure 10-9.

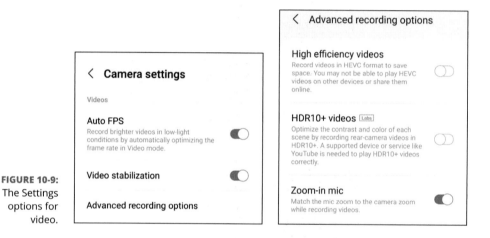

Here's more on what these three options mean:

>> **Auto FPS:** When you're shooting in low light conditions (when it's kind of dark), things can get dicey and the video may not look right. To address this, the S22 can reduce the number of frames to soak in what light there is and make the video appear clearer.

>> **Video Stabilization:** This is the same option that's on the viewfinder. Trust me. Leave it on.

>> **Advanced Recording Options:** The Samsung Galaxy S22 can let you get really carried away with these options. You can, in spite of having gobs of storage if you upgraded, choose to use the high-efficiency option. You can also use HDR10+, which plays with contrast levels. Unless you know that this option is for you, you probably can skip it. The option you may want is the zoom-in mic. This will guess who is the most important person in the frame that you're following, and will have the microphone pick up their voice out of the mix. Pretty cool.

# Messing with the AR Doodle option

When you tap the squiggly line in the corner of the viewfinder, you can have some fun. This brings up the screen shown in Figure 10-10.

**FIGURE 10-10:** The start screen for AR Doodle.

Here, you shoot video for about a minute. Then you're offered some writing options. For example, feel free to draw a Snidely Whiplash mustache on your dad. Now that mustache will be following him around in the viewfinder.

Silly, but what good is all this elaborate technology if you can't have some fun?!

# Chapter **11**

# Playing Games

G ames are the most popular kind of download for smartphones of all kinds. In spite of the focus on business productivity, socializing, and making your life simpler, games outpace all other app downloads. To this point, the electronic gaming industry has larger revenues than the movie industry — and has for several years!

We could have a lively and intellectually stimulating debate on the merits of games versus applications. For the purposes of this book, the differences between games and apps are as follows:

» If people like a game, they tend to play it for a while, maybe for a few weeks or even months, and then stops using it. A person who likes an app tends to keep on using it.

» Games tend to use more of the graphical capabilities of your phone.

» People who use their phones for games tend to like to try a wide range of games.

The fact of the matter is that your Samsung Galaxy S22, with its large Super AMOLED screen and beefy graphics processing unit, makes Android-based games more fun. And because you already have one, maybe you should take a break and concentrate on having fun!

# The Play Store Games Category

Chapter 8 introduces the Play Store, shown in Figure 11-1. The top level splits offerings into a few categories: Games, Apps, Movies & TV, and Books.

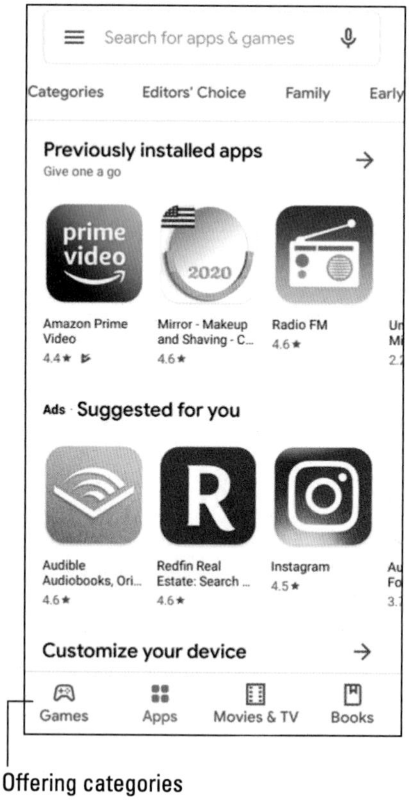

**FIGURE 11-1:** The Play Store Home screen.

Offering categories

We want games. Games that test our skills; games that are fun; games that are cute; games that immerse us in an alternate universe! To get there, tap on the Games button!

This brings up the Games page as shown in Figure 11-2.

This section of the store has nothing but games. This section includes everything from simple puzzles to simulated violence. All games involve various combinations of intellect, skill (either cognitive or motor), and role-playing. Let's do it.

**FIGURE 11-2:**
The Games link
on the Google
Play Home
screen.

Offering Categories

## The Games Home screen

If you scroll around this screen, you see many suggested games. This is shown in panorama in Figure 11-3.

If you aren't sure what games you might like to try, don't worry: There are lots of options. As you can see, the Games Home screen makes lots of suggestions. Each row takes a different perspective on helping you find a new game. A few of these are board games, strategy games, and action games. They also include games that allow you to play offline without Wi-Fi.

Another approach is to choose the Categories options, shown partially in the top right in Figure 11-2. This will bring up the game categories shown in Figure 11-4.

FIGURE 11-3:
The Games
Home screen
in Panorama.

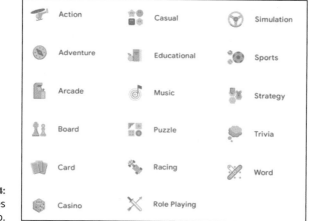

| | | | |
|---|---|---|
| Action | Casual | Simulation |
| Adventure | Educational | Sports |
| Arcade | Music | Strategy |
| Board | Puzzle | Trivia |
| Card | Racing | Word |
| Casino | Role Playing | |

FIGURE 11-4:
The Games
Categories tab.

## The Games Categories tab

In the Play Store, games are divided into the following genres:

>> **Action:** Games that involve shooting projectiles that can range from marsh-mallows to bullets to antiballistic missiles. They also involve fighting games with every level of gore possible.

>> **Adventure:** Games that take you to virtual worlds where you search for treasure and/or fight evil. Zombies and vampires are traditional evildoers.

>> **Arcade:** Game room and bar favorites.

>> **Board:** Versions of familiar (and some not-so-familiar) board games.

>> **Card:** All the standard card games are here.

>> **Casino:** Simulations of gambling games; no real money changes hands.

>> **Casual:** Games that you can easily pick up and put aside (unless you have an addictive personality).

>> **Educational:** Enjoyable games that also offer users enhanced skills or information.

>> **Music:** Includes a wide range of games that involve music in one way or another. These games may include trivia, educational games involving learning music, or sing-along songs for kids.

>> **Puzzle:** Includes games like Sudoku, word search, and Trivial Pursuit.

>> **Racing:** Cars, go-karts, snowboards, jet skis, biplanes, jets, or spacecraft competing with one another.

>> **Role Playing:** In a virtual world, become a different version of who you are in real life, be it for better or worse.

>> **Simulation:** Rather than live in the virtual world of some game designer, create and manage your own virtual world.

>> **Sports:** Electronic interpretations of real-world activities that incorporate some of the skill or strategy elements of the original game; vary based upon the level of detail.

>> **Strategy:** Emphasize decision-making skills, like chess; a variety of games with varying levels of complexity and agreement with reality.

>> **Trivia:** A variety of games that reward you if you know things like the name of the family dog from the TV show *My Three Sons.* Its name was Tramp, but you knew that already.

>> **Word:** Games that are universally popular, such as Scrabble.

**TIP**

Many games appear in more than one category.

Each game has a Description page. It's similar to the Description page for apps, but it emphasizes different attributes. Figure 11-5 is an example of a typical Description page.

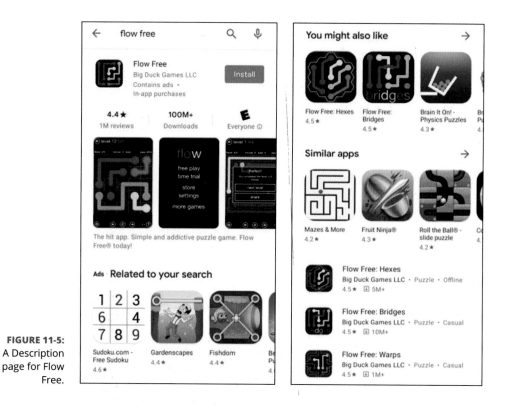

FIGURE 11-5:
A Description
page for Flow
Free.

When you're in a category that looks promising, look for these road signs to help you check out and narrow your choices among similar titles:

>> **Ratings/Comments:** Gamers love to exalt good games and bash bad ones. The comments for the game shown in Figure 11-5 are complimentary, and the overall ranking next to the game name at the top suggests that many others are favorable.

>> **About This Game:** This tells you the basic idea behind the game.

>> **What's New:** This section tells what capabilities have been added since the previous release. This is relevant if you have an earlier version of this game.

>> **Reviews:** Here is where existing users get to vent their spleen if they do not like the game or brag about how smart they are for buying it ahead of you. The comments are anonymous, include the date the comment was left, and

tell you the kind of device the commenter used. There can be applications that lag on some older devices. However, you have the Galaxy S22, which has the best of everything (for now).

>> **More Games by Developer:** If you have a positive experience with a given game, you may want to check that developer's other games. The More By section makes it easier for you to find these other titles.

>> **Users Also Viewed/Users Also Installed:** This shows you the other apps that other people who downloaded this app have viewed or downloaded. These are some apps that you may want to check out.

>> **Price:** As a tie-breaker among similar titles, a slightly higher price is a final indication of a superior game. And because you're only talking a few pennies, price isn't usually a big deal.

# Leaving Feedback on Games

For apps in general, and games in particular, the Play Store is a free market. When you come in to the Play Store, your best path to finding a good purchase is to read the reviews of those who have gone before you. Although more than a million users have commented on Angry Birds, most games do not have that kind of following.

One could argue that your opinion would not move the overall rating for a frequently reviewed game like Angry Birds. The same cannot be said for other games.

One of the new games is Santa City Run from rayan studio. The game description is shown in Figure 11-6.

REMEMBER

A Description page, before you download it to your phone, will show either the Install button or the price of the game; the feedback areas will be grayed out. The Description page after you download the game to your phone will offer the options to Open or Uninstall, and the feedback areas will be active.

**FIGURE 11-6:**
A game
description for
Santa City Run.

In this case, Santa City Run has not been reviewed at all. Your opinion matters more for this game than for the heavily reviewed games. After you've downloaded and played a game, you can help make the system work by providing your own review. This section reviews the process, starting from the first screen of the Play Store (refer to Figure 11-1):

1. **Tap the Menu icon.**

   This brings up a pop-up menu like the one shown in Figure 11-7.

2. **Tap Manage Apps & Device.**

   You're taken to the screen shown in Figure 11-8.

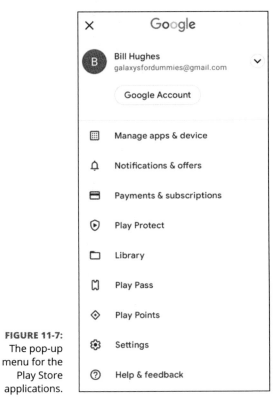

**FIGURE 11-7:**
The pop-up
menu for the
Play Store
applications.

**FIGURE 11-8:**
Check out your
downloads.

## 3. Tap the Manage link.

You see the apps and games that you've downloaded, as shown in Figure 11-9. The Play Store doesn't distinguish between games and apps in this menu when showing you the list of installed apps. They're all in the same list. However, you have the option to scroll to the right and see a link that says Games.

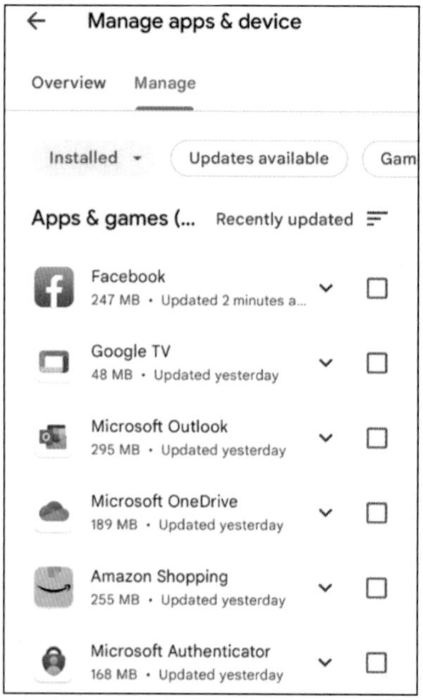

FIGURE 11-9:
The Manage screen for apps and games.

## 4. Tap the Games link.

You see just the games, as shown in Figure 11-10, making it easier to manage your games.

## 5. Tap the game you'd like to leave feedback for.

Tapping the title of the game normally brings up the game description similar to what is shown in Figures 11-5 and 11-6. After you've downloaded a game, however, a Rate This App section appears that lets you leave feedback. See Figure 11-11 to see this section for Angry Birds 2.

## 6. Tap the stars on the screen.

This brings up a pop-up screen, as shown on the left of Figure 11-12.

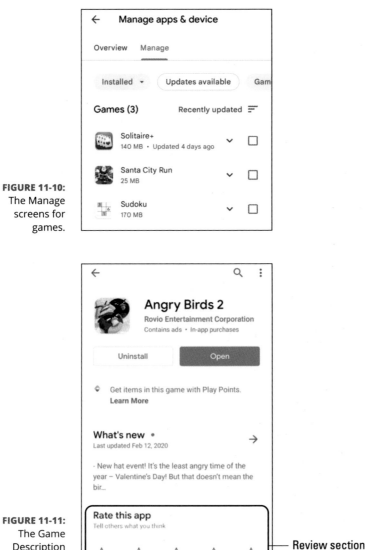

**FIGURE 11-10:**
The Manage screens for games.

**FIGURE 11-11:**
The Game Description page with space for feedback.

Review section

**7.** **Tap the number of stars you believe this game deserves.**

The stars are between one and five. You're asked to answer some other questions that are shown in Figure 11-12, including making any comments on the last pop-up.

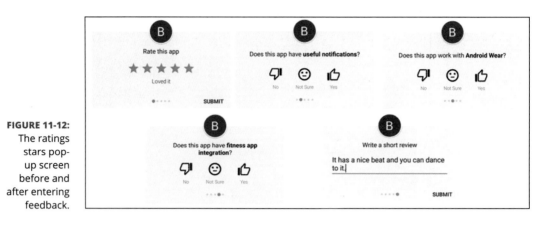

**FIGURE 11-12:**
The ratings
stars pop-
up screen
before and
after entering
feedback.

8. **When you're done, tap Submit.**

   Your comments are sent to the Play Store for everyone to see. For the sake of the system, make sure your comments are accurate!

**TIP**

Free apps are great, but don't be afraid to buy any apps that you think you'll use frequently. Games usually cost very little, and the extra features may be worth it. Some people (including me) have an irrational resistance to paying $1.99 monthly for something they use all the time. Frankly, this is a little silly. Let's all be rational and be willing to pay a little bit for the services we use.

# Chapter **12**
# Playing Music and Videos

Smartphones have built-in digital music and video players. There was a time not too long ago that having a dedicated music or video player was the way to go. However, having a single device that you can use as a phone *and* as a source of entertainment is more convenient because then you need only one device rather than two, and you eliminate extra cords for charging and separate headphones for listening. Your Samsung Galaxy S22 is no exception. You can play digital music files and podcasts all day and all night on your phone.

The audio and video technology embedded in your phone is truly among the best there is. This is not just adequate technology; this is state-of-the-art, over-the-top, best-that-money-can-buy technology. You can choose to pursue your level of enjoyment that is convenient, or you can buy complementary equipment that can fully leverage the capabilities within the S22. This is your choice, and no one will judge you (although I may be envious if you buy some of the high-end equipment).

In addition, by virtue of the Super AMOLED screen on your Galaxy S22 smartphone, your phone makes for an excellent handheld video player. Just listen on the headset and watch on the screen, whether you have a short music video or a full-length movie. You may look at your phone and wonder how good watching a video can be with a relatively small screen, but once you start, the screen is so good, you become immersed in no time.

To boot, your Galaxy S22 comes with applications for downloading and listening to music as well as downloading and watching videos. These apps are very straightforward, especially if you've ever used a CD or a DVD player.

# Getting Ready to Be Entertained

Regardless of the model phone you have, the particular app you use for entertainment, and whether you're listening to audio or watching video, here's where I cover some common considerations up front.

The first is the use of headsets. Yeah, you used *headphones*, but you probably want to use a *headset* with your phone. The vocabulary is more than just semantics; a headset has headphones plus a microphone on the cable. This makes it easier for when you get a call while listening to music or watching a movie.

Your phone sees an incoming call and will pause the music or movie and show you the caller ID. If you answer it and you have a headset, you can be reasonably assured that the person on the other line will hear you loud and clear. If you have headphones, the phone will use the built-in microphone for your voice while you hear the caller on your headphones. My advice is to test the sound quality using a headphone and judge for yourself whether you want to use a headset or headphones.

Next, I explore speakers. I touch on speakers in Chapter 3 when talking about Bluetooth pairing and in Chapter 16 when I talk about making the phone your own. As you may guess, having a solid strategy for speakers is important to get the most out of your Galaxy S22.

Then you need to know about connecting your Galaxy S22 phone to a television and/or stereo. After I talk about that, I cover the issue of licensing multimedia material.

Finally, I explore storing all these options on your phone for your offline enjoyment. You can have gobs of storage at your disposal if you want. I explore what the term "gobs" really means in terms of audio and video storage.

## Choosing your headset

You can use wired or wireless (Bluetooth) headsets with your Samsung Galaxy S22 phone. Wired headsets are less expensive than Bluetooth headsets, and, of course, wired headsets don't need charging, as do the Bluetooth headsets.

On the other hand, you lose freedom of mobility if you're tangled up in wires. In addition, the battery in Bluetooth headsets lasts much longer than the battery in your phone. However, if your headset happens to be running low and it can be charged wirelessly, you can use the PowerShare feature of your S22 to extend the life of your headset until you can recharge everything.

# Wired headsets

At the bottom of your Galaxy S22 phone, you won't see a regular headset jack. What you do see is the USB-C port. There are two choices. The first choice is to get headsets that have a USB-C connector. Your phone may come with a wired headset that has a USB-C connector like those shown in Figure 12-1 on the left. In that case, just plug it in to use the device.

Headset with USB-C Connector

3.5 mm to USB-C Connector

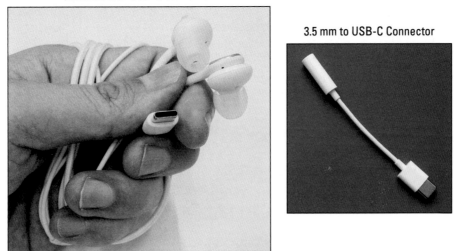

Left: Heng Lim/Shutterstock; Right: vta_photo/Shutterstock

**FIGURE 12-1:** A wired headset with a USB-C connector and a 3.5mm to USB-C connector.

The other option is to get a connector that allows you to plug in your existing headphone/headset. The image on the right in Figure 12-1 is an example of one of these connectors (also called a *dongle*). The retail cost is about $20, but you can find them online for about $10.

When you do plug in, you'll hear the audio, but the person on the other end of the call may not hear you as well because the headphones don't come with a microphone. In such a case, your phone tries to use the built-in mic as a speakerphone. Depending upon the ambient noise conditions, it may work fine or sound awful. Of course, you can always ask the person you're talking to whether they can hear you.

Some people dislike ear buds. You can obtain other styles at a number of retail franchises that offer options, including:

>> Around-the-ear headphones that place the speakers on the ear and are held in place with a clip

>> A behind-the-neck band that holds around-the-ear headphones in place

>> An over-the-head band that places the headphones on the ear

**WARNING**

The laws in some regions prohibit the use of headphones while driving. Correcting the officer and explaining that these are really "headsets" and not "headphones" won't help your case if you're pulled over. Even if not explicitly illegal in an area, it's still a bad idea to play music in both ears at a volume that inhibits your ability to hear warnings while driving.

**WARNING**

Ear buds can have a greater chance of causing ear damage if the volume is too loud than other options. The close proximity to your eardrum is the culprit. There are probably warnings on the ear bud instructions, but I wanted to amplify this information (<harhar>).

In any case, give yourself some time to get used to any new headset. There is often an adjustment period while you get used to having a foreign object in or around your ear.

## Stereo Bluetooth headsets

The other option is to use a stereo Bluetooth headset. Figure 12-2 shows a few options.

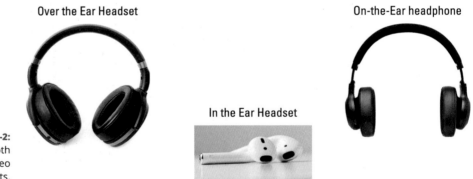

Over the Ear Headset

On-the-Ear headphone

In the Ear Headset

**FIGURE 12-2:**
Bluetooth stereo headsets.

Left: timquo/Shutterstock; Center: alexanderon/Shutterstock; Right: Photoongraphy/Shutterstock

You need to consider a few important options before you put down your money and buy one. First, you have a choice of the earpieces being "on" the ear, "over" the ear, or "in" the ear.

The over-the-ear option tends to eliminate outside noise a little better because your ears are surrounded by the cushions. This can also get hot if you wear your headset for an extended time. On the other hand, some over-the-ear headsets have noise cancelling, which is nice in noisy areas, like airplanes. The on-the-ear headphone is good for when you have your headphones on for a long time.

The third option for headsets includes those that insert into your ear canal. The ear buds in Figure 12-2 have extensions that hang out of your ear, but some are self-contained in a very small package. When you insert them into your ear canal, they just stay there (if they don't, there are a few other silicone ear pieces that you can use to get them to fit). There are no wires or plugs. Having this kind of headset in your ear makes it less likely to shift or fall out than the other options. The in-the-ear headset shown in Figure 12-2 is well-suited for those who want to listen to music and have the ability to take calls as they exercise. How neat is that?

In each of these cases, a stereo Bluetooth headset is paired the same way as any other Bluetooth headset. (Read how to do this in Chapter 3.) When your Galaxy S22 phone and the headset connect, the phone recognizes that the headset operates in stereo when you're listening to music or videos. When you open the case in which you store the ear pieces, not only will the Galaxy Buds pair, but they'll bring up a screen called Galaxy Gear to let you fine-tune the sound quality. It's very slick!

## Choosing your Bluetooth speaker

In the last few years, developers released a flurry of products known as *Bluetooth speakers*. These speakers include a range of options, some of which are a very small and convenient. Others are designed to offer excellent audio quality. Some of these speakers look cool or are very small but have poor sound quality. Check them out before your buy.

Although these speakers (which come in a range of sizes) are not as portable as the Bluetooth headset — they're a little difficult to use as you're walking down the street — they're usually pretty easy to take with you and set up when you're at a desk or in someone's living room. They also do not need a cable to make a connection and are always ready to go.

Audio Pro's A36 Bluetooth speaker (see Figure 12-3) is an excellent example of a high-quality Bluetooth speaker. Its list price is about $900.

If you like high-quality sound, this is the quality of Bluetooth speaker that you'd want to get and your S22 is up to the task. As you may have noticed, it looks very sleek. On the other hand, if you're just looking for background enjoyment, you can get a more modest Bluetooth speaker for a fraction of that price.

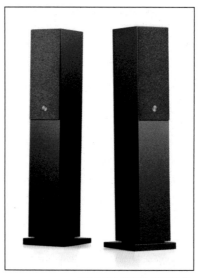

*Photograph courtesy of Audio Pro*

## Connecting to your stereo

Although being able to listen to your music on the move is convenient, it's also nice to be able to listen to your music collection on your home or car stereo. Your Galaxy S22 phone presents your stereo with a nearly perfect version of what went in. The sound quality that comes out is limited only by the quality of your stereo.

In addition, you can play the music files and playlists stored on your phone, which can be more convenient than playing CDs. If your stereo receiver is older, you have two choices:

» One setup involves plugging a 3.5mm-to-USB-C connector into the 3.5mm jack available on newer stereos. You can also acquire a cable with the male 3.5mm-to-a-male USB-C connector, but these are just becoming available as of this writing.

When you play the music as you do through a headset, it will play through your stereo. Although each stereo system is unique, the correct setting for the selector knob is AUX.

» The other option is to acquire a Bluetooth receiver and connect that to your stereo. Be sure it works at least with Bluetooth version 4.2 for the best sound. Your S22 has version 5.2, but version 4.2 is when things got really good, and version 5.2 happily works with version 4.2. The Bluetooth receiver will connect with a 3.5mm jack or to an unused red and white plug pair, say, for the old cassette tape player you no longer have. This setup lets you walk around with your phone in your house. As long as you don't stray more than 30 feet from your stereo, the connection works great!

Newer stereo receivers recognize that cellular phones are a great place to access music. These receivers support a Bluetooth connection to your phone. If you are in the market for a new receiver, make sure that you get one with Bluetooth capability. In either case, you will be entertained. Enjoy yourself.

# Licensing Your Multimedia Files

It's really quite simple: You need to pay the artist if you're going to listen to music or watch video with integrity. Many low-cost options are suitable for any budget. Depending upon how much you plan to listen to music or podcasts, or watch videos, you can figure out what's the best deal.

**WARNING**

Stealing music or videos is uncool. Although it might be technically possible to play pirated music and videos on your phone, it's stealing. Don't do it. If your financial circumstances do not allow you to afford to pay for your music, I suggest you listen to Internet Radio. With just a little work on your part, you can get free unlimited music and at least a low monthly fee if you want no advertising.

You can buy or lease music, podcasts, or videos. In most cases, you pay for them with a credit card. And depending upon your cellular carrier, you might be allowed to pay for them on your monthly cellular bill.

## Listening up on licensing

Here are the three primary licensing options available for music files and podcasts:

>> **By the track:** Pay for each song individually. Buying a typical song costs about 79 cents to $1.29. Podcasts, which are frequently used for speeches or lectures, can vary dramatically in price.

>> **By the album:** Buying an album isn't a holdover from the days before digital music. Music artists and producers create albums with an organization of songs that offer a consistent feeling or mood. Although many music-playing apps allow you to assemble your own playlist, an album is created by professionals. In addition, buying a full album is often less expensive than on a per-song basis. You can get multiple songs for $8 to $12.

>> **With a monthly pass:** The last option for buying audio files is the monthly pass. For about $15 per month, you can download as much music as you want from the library of the service provider.

If you let your subscription to your monthly pass provider lapse, you won't be able to listen to the music from this library. This music is *streamed* through the Internet and not stored on your phone.

In addition to full access to the music library, some music library providers offer special services to introduce you to music that's similar to what you've been playing. These services are a very convenient way to learn about new music. If you have even a small interest in expanding your music repertoire, these services are an easy way to do it.

Whether buying or renting is most economical depends on your listening/viewing habits. If you don't plan to buy much, or you know specifically what you want, you may save some money by paying for all your files individually. If you're not really sure what you want, or you like a huge variety of music, paying for monthly access might make better sense for you.

## Licensing for videos

The two primary licensing options available for videos are:

>> **Rental:** This option is similar to renting a video from a store. You can view the video as many times as you like within 24 hours from the start of the first play. In addition, the first play must begin within a defined period, such as a week, of your downloading it. Most movies are in the $3 to $5 range.

>> **Purchase:** You have a license to view the file as frequently as you want, for as long as you want. The purchase cost can be as low as $12, but is more typically in the $15 range.

At the moment, there are no sources for mainstream Hollywood films that allow you to buy a monthly subscription and give you unlimited access to a film library. This can change at any time, so watch for announcements.

# Using the Full Capacity of the Memory in Your Phone

Your Galaxy S22 comes with at least 128GB of storage. Okay. What does that really mean?

**TIP**

Prior models of the Galaxy offered you the option to supplement the memory installed in your phone with an SD card, but this is no longer an option. For this reason, you need to select a phone with enough memory when you first buy the phone. The least amount of memory you can buy is 128GB. You can buy a model with 256GB. The S22 Ultra offers the option for 512GB. Choose wisely. The Android operating system is stored on some of that storage space. Plus, the apps that you downloaded are stored here. Then, there are the files that you create and use. After it is all said and done, you probably have at least 80GB available for the storage of photos, music, and videos. So how much is that, and do you need more?

The correct answer is, "Who knows?" That is about as worthless of an answer as you will find anywhere in this book. However, I can tell you how to change that response to "It doesn't matter."

The easiest way to describe it is by figuring out how many photos, how many songs, and how many movies fit into 1GB.

>> **300 images:** You can take about 300 images from the camera on your phone at resolution at the default settings.

>> **Ten hours of music:** If you buy an album, it will take somewhat less than a tenth of a gigabyte. The length of albums was set based upon what could be reliably imprinted on a vinyl record. That works out to be about 45 minutes and maybe a dozen songs.

>> **About one standard definition (TV-quality) movie:** Estimating the size of videos can get really variable. TV quality video is a waste of that beautiful screen. However, an hour of video that uses every pixel of resolution will take about 4GB.

Unfortunately, at this point, you need to do a little bit of math. Say that I want to watch every minute of all 529 episodes of *The Simpsons*. This would take 194 hours or about 8 days. In this case, you would want the 256GB option for about $50 more. D'oh!

You could binge-watch all the episodes of the series *Breaking Bad* in high definition, but then you would need to delete *The Simpsons*. Sorry.

If you are more into music than videos, you could store over 5,000 hours of music on your phone with 256GB. You could start playing music January 1 at midnight and not repeat a single track until July 27.

Keep in mind that these are rough estimates. These calculations assume there are minor attempts to reduce the size of these files. That can have a huge impact on storage needs. Let's just say that you can put a combination of images, music, and video that will keep you entertained for a very, very long time.

# Enjoying Basic Multimedia Capabilities

Regardless of the version of your Galaxy S22, some basic multimedia capabilities are common across the different phones. Figure 12-4 shows the Music and Video applications that come with all Galaxy S22 phones.

**FIGURE 12-4:**
The multimedia apps on the Galaxy S22.

YouTube Music

Google TV

Your phone comes with these apps preloaded, and you might have other multimedia apps as well, depending upon your carrier. Also, some sources of music and video, such as Netflix, have their own embedded player.

## Grooving with the YouTube Music app

The YouTube Music app allows you to play music and audio files. The first step is to obtain music and audio files for your phone.

Some ways to acquire music and/or recordings for your phone are:

>> Buy and download tracks from an online music store.

>> Receive them as attachments via email or text message.

>> Receive them from another device connected with a Bluetooth link.

>> Record them on your phone.

### Buying from an online music store

The most straightforward method of getting music on your phone is from an online music store. You can download a wide variety of music from dozens of mainstream online music stores when you're a subscriber. The Play Store is an option. In addition to apps, it has music and video. Other well-known sites include Spotify and Amazon Music. In some cases, such as the Google Play Store or Amazon, you may already be a subscriber and not know it. When the music starts playing, you have your answer. Otherwise, you may need to provide an email address and credit card number to get the ball rolling.

In addition to the mainstream choices, many more specialty or "boutique" stores provide more differentiated offerings than you can get from the mass-market stores. For example, MAQAM offers Middle Eastern music (www.maqammp3.com).

The details for acquiring music among the online stores vary from store to store. Ultimately, there are more similarities than differences. As an example of when you know what you want, what you really, really want, here's how to find the song "Wannabe" by the Spice Girls. I'm using Amazon Music. If you don't have Amazon Music in your Applications list, you would start by loading that app on your phone, as I describe how to do in Chapter 8. After entering your Amazon sign-in information, you see the Amazon Music Home screen, shown on the right in Figure 12-5.

Amazon Music Sign-In Screen          Amazon MP3 Music Store Home Screen

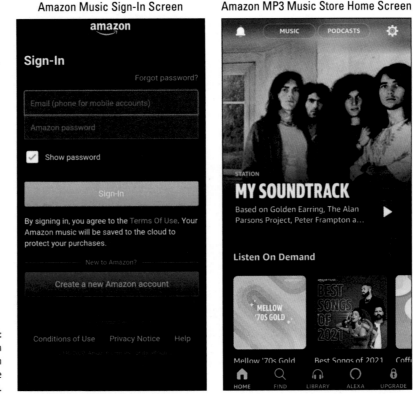

FIGURE 12-5:
The Amazon
Music login
and Home
screens.

From here, you can search for music by album, song, or music genre.

**TIP**

Many of the music apps try very hard to get you to sign up for their monthly service. This can be annoying if you want to dip your toe in the water before you dive in. As a newcomer, knowing which music service will best serve your needs is difficult, even if they offer you a 30- or 90-day free trial. Be patient and keep your options open until you're comfortable. Or, take them up on their offers, but just remember to cancel a service if you aren't using it. They tend to assume that you want their service even if you've never used it after the first few days, and they'll keep charging you until you cancel.

Now to search for the song you want:

1. **Enter the relevant search information in the Amazon Music Find field.**

   In this case, I'm searching for "Wannabe" by the Spice Girls. The result of the search for songs looks like Figure 12-6.

   The search results come up with all kinds of options, including: albums, individual tracks, similar songs, karaoke versions, and other songs from the same artist. Be ready for these options.

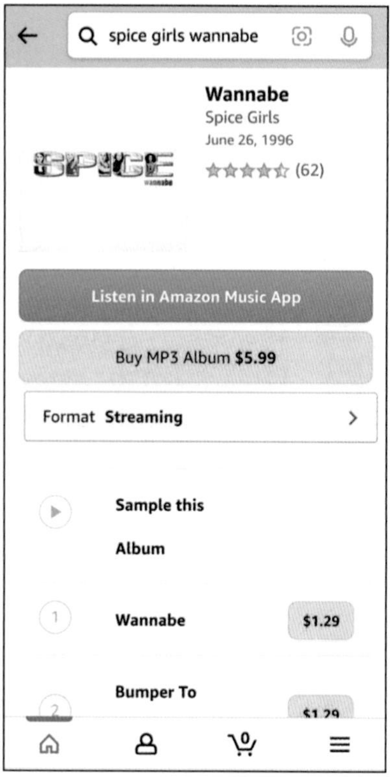

**FIGURE 12-6:**
Search results for a song at the Amazon Music store.

2. **To buy the track, tap the yellow box where it says $1.29.**

   A pop-up like the one shown in Figure 12-7 appears.

FIGURE 12-7:
Tap to buy.

**3.** **To buy the song, tap the box that says, "Pay with USD."**

The music store will start playing this song followed by a list of music that is dominated by Spice Girls songs. They also throw in similar songs. This is called a "Station." In this case, you get the screen shown in Figure 12-8.

When you open the music player, it's ready for you to play anytime you want to party like it's 1996!

## Receiving music as an attachment

As long as you comply with your license agreement, you can email or text a music file as an attachment. Simply send yourself an email from your PC with the desired music file. You then open the email or text on your phone and save the file in the library of your Music app.

TECHNICAL
STUFF

Your phone can play music files that come in any of the following formats: FLAC, WAV, Vorbis, MP3, AAC, AAC+, eAAC+, WMA, AMR-NB, AMR-WB, MID, AC3, and XMF.

**FIGURE 12-8:**
The Amazon
Music store
starts playing
your tune.

## Recording sounds on your phone

No one else might think your kids' rendition of "Happy Birthday" is anything special, but you probably treasure it. In fact, many sound recording apps are available on your phone.

Some are basic and turn your phone into a basic voice recorder. Others can alter your voice. (I do not want to know why.) Others are meant to allow you to surreptitiously record voice conversations. Some can specifically record telephone conversations.

**TIP**

Be aware of privacy laws in your area! Some jurisdictions require only one party in a conversation to give permission to record a conversation. Other jurisdictions require positive consent from all parties. While rare, there have been cases where first-time offenders have spent long stints in jail for recording what was perfectly legal in another state!

In general, a simple record button creates a sound file when you stop recording. The sound quality may not be the best, but what you record can be just as important or entertaining as what you buy commercially. Your phone treats all audio files the same and all are playable on your YouTube Music app.

# Playing downloaded music

To play your music, tap the YouTube Music app icon (refer to Figure 12-4) to open the music playing application.

Okay. Figure 12-9 shows your Home screen after your get through the initial marketing welcome. Sometimes this app can be too enthusiastic in welcoming you.

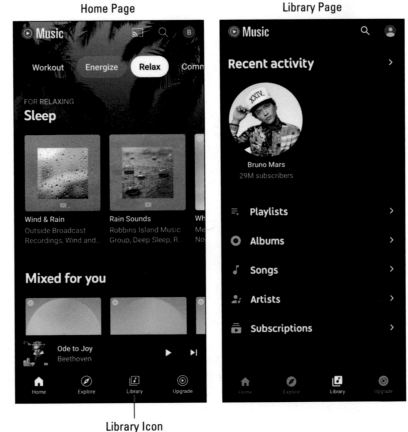

Home Page                    Library Page

**FIGURE 12-9:**
The Home
screen and
Library for
the YouTube
Music app.

Library Icon

When you tap the Library icon, you see the screen to the right. It sorts your music files into a number of categories. You can select the category you want to use by tapping on the link.

The categories include

>> **Playlists:** Some digital music stores bundle songs into playlists, such as Top Hits from the '50s. You can also create your own playlists for groups of songs that are meaningful to you.

>> **Albums:** Tapping this category places all your songs into an album with which the song is associated. When you tap the album, you see all the songs you've purchased, whether one song or all the songs from that album.

>> **Songs:** This lists all your song files in alphabetic order.

>> **Artists:** This category lists all songs from all the albums from a given artist.

>> **Subscriptions:** You can upgrade and then subscribe to the specific music genres you like; they're called stations.

TIP

Many music services have an option called *stations.* These are kind of like radio stations that follow a particular format. You put in a group name, like Genesis or Van Halen, and the app will play songs within that genre that may or may not be currently stored on your phone. Figure 12-10 shows a more recent example when I searched on Bruno Mars.

**FIGURE 12-10:**
Bruno Mars
Radio Station
within the
YouTube
Music app.

These categories are useful when you have a large number of files. To play a song, an album, or a genre, open that category and tap the song, playlist, album, artist, or genre, and the song will start playing.

## Adding songs as ringtones and alarms

Here's how to add a song as a ringtone or alarm. The first step is to open a contact. A generic contact in Edit mode is seen in Figure 12-11. (Refer to Chapter 6 if you have any questions about contacts.)

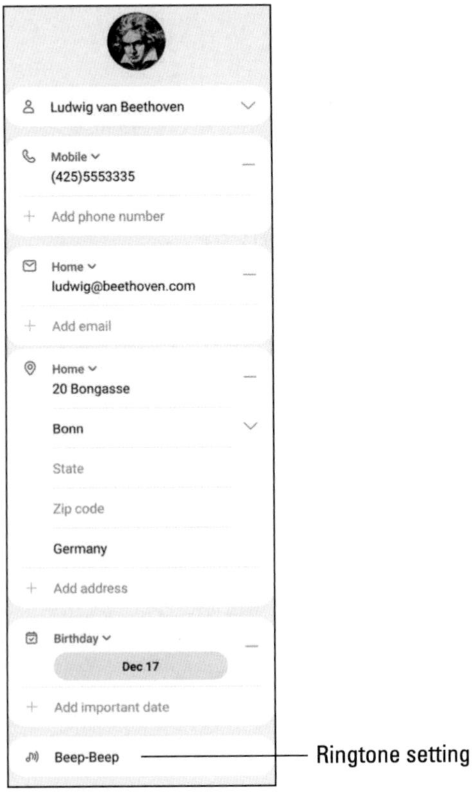

**FIGURE 12-11:**
A typical
contact for
a Baroque
composer.

Ringtone setting

Follow these steps:

**1. Tap the Default Ringtone link.**

This will bring up a list of ringtones (shown in Figure 12-12).

A quick scan finds that *Ode to Joy* is not among the options that come with your phone. To use a music file as a ringtone, find the Add button at the top.

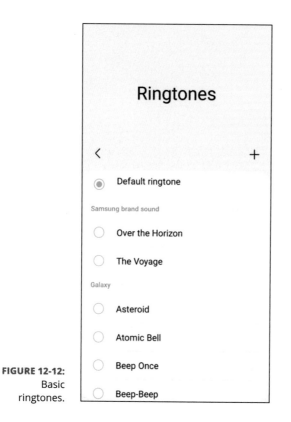

**FIGURE 12-12:**
Basic
ringtones.

**2. Tap Add.**

This brings up the option to bring up the Sound Picker App. This app shows all the music files on your phone, as shown in Figure 12-13.

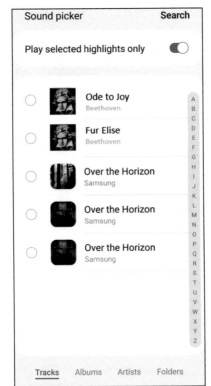

**FIGURE 12-13:**
The track
selections.

3. **Highlight the song you want and tap OK.**

From now on, when you hear this song, you know it will be your friend Ludwig. Figure 12-14 shows that this is now Ludwig's ringtone.

## Jamming to Internet Radio

If you have not tried Internet Radio, you should definitely consider it. The basic idea is that you don't need to be near the station to receive its broadcasts. For example, I can be in Seattle and enjoy WXRT, which offers Chicago's Finest Rock.

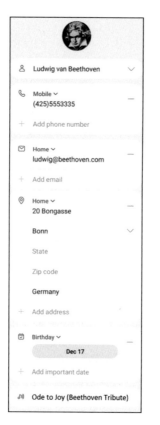

**FIGURE 12-14:**
The ringtone
selection.

The good news is that you can explore your options. Figure 12-15 shows some of the 9,000 Internet Radio apps in the Play Store. Pandora, IHeartRadio, and Slacker Radio are three of the best-known services of this type; one or the other may have been preinstalled on your phone. (Actually, you may find four or more Internet Radio options preinstalled on your phone!)

These apps are a great way to learn about new songs and groups that may appeal to you. The service streams music to your phone for you to enjoy. You can buy the track if you choose.

**WARNING**

Streaming audio files can use a large amount of data over time. This may be no problem if you have an unlimited (or even large) data-service plan. Otherwise, your "free" Internet Radio service can wind up costing you a lot. You're best off using Wi-Fi.

**TIP**

If you find yourself bored, do not hesitate to switch Internet Radio providers and use the same band for the radio station. This will almost always bring up new songs.

**FIGURE 12-15:**
Some Internet
Radio options
in the Play
Store.

## Looking at your video options

The YouTube Music app allows you to play music files. Similarly, you can use the Google TV app to play video options. Google TV is in your Applications list and might even be on your home page. In most ways, playing videos is the same as playing audio with some exceptions:

» Many people prefer to buy music, but renting is more typical for videos.

» Video files are usually, but not always, larger.

Otherwise, as with music files, you can acquire videos for your phone from an online video store — and you need to have an account and pay for the use. In addition, you can download video files to your phone, and Google TV will play them like a DVD player.

There is a great selection of videos on the Google Play Store and Amazon Prime Video. Each of these has great video selections that you can rent or buy. Figure 12-16 shows the Home screens for the Google Play Store and Amazon Prime Video.

Play Store Video Home Page          Amazon Prime Video Home Page

**FIGURE 12-16:**
Home screens
for Google
Play Store and
Amazon Prime
Video.

## Using the three screens

If you have a subscription to Amazon Video, I hereby grant you permission to watch any and all of the Amazon Video options on your Galaxy S22 (once you sign in and comply with all the terms and conditions set forth by Amazon). Once you install the Amazon Prime Video app from the Play Store, sign in with the email and password associated with your account, and all the content is there for you to stream. It is that simple. If you don't believe me, give it a try.

If you take a look at the Amazon Prime Video home page on the Internet, seen in Figure 12-17, it shows your options for access to the content to which you subscribe. The original term for this was called serving the *three screens.* Three screens referred to in the strategy included your television at home, your PC or laptop, and your smartphone.

The idea is that you get one subscription and have access to the same content and, importantly, can pick up where you left off. So if you're watching a video on your television, you can pick up where you left off on your smartphone.

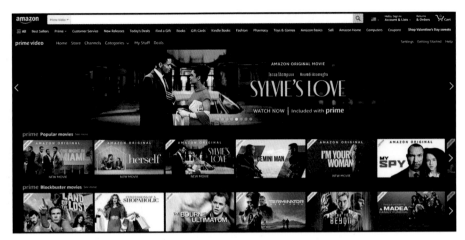

**FIGURE 12-17:**
The Amazon
Prime Video
home page.

Amazon Prime Video is taking this one step further to ensure as many of its sub-scribers as possible have access. If you have a Smart TV that has an Internet con-nection, the chances are that Amazon Prime Video will run on the TV. If you have an old and/or a dumb TV, you can get Amazon through streaming media players, game consoles, set-top boxes, or Blu-ray players.

Amazon is not the only organization to do this. Many cable companies offer this kind of solution, as do many of the video subscription services.

The mainstream video services compete with having a broad range within their libraries that seek to appeal to as many customers as possible. Keep in mind that there are specialty video providers that offer curated videos for their subscribers.

For example, TeacherTube is a site dedicated to K-12 education, as seen in Figure 12-18. If you continue down this path further, there are a great number of options for online education. Many of these sites do not consider themselves to be video aggregators, but that's exactly what happens when they take recorded lectures and provide them to students.

The best-known online education service is the University of Phoenix. There are dozens more online universities.

Education is just one curated video service. Others exist for videos of Bollywood movies, British sitcoms, Portuguese game shows, and many other art forms.

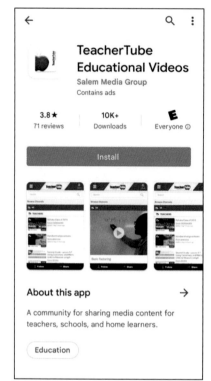

**FIGURE 12-18:**
The TeacherTube
app.

## Viewing your own videos

In Chapter 10, I cover how to use the digital camcorder on your phone. You can watch any video you've shot on your phone. From the Google Play app, scroll over to the Personal Video section.

Your phone can show the following video formats: MPEG-4, WMV, AVI/DivX, MKV, and FLV.

**TECHNICAL STUFF**

To play your video, simply tap the name of the file. The app begins showing the video in landscape orientation. The controls that pop up when you tap the screen are similar to the controls of a DVD player.

# 5

# Getting Down to Business

Download your calendars to your phone and upload new events to your electronic calendar.

Use Google Maps and the GPS in your phone to get you there and back.

Pay with Samsung Pay.

# Chapter **13**
# Using the Calendar

You might fall in love with your Galaxy S22 phone so much that you want to ask it out on a date. And speaking of dates, let's talk about your phone's calendar. The Galaxy S22 phone calendar functions are powerful, and they can make your life easier. With just a few taps, you can bring all your electronic calendars together to keep your life synchronized.

In this chapter, I show you how to set up the calendar that comes with your phone, which might be all you need. The odds are, though, that you have calendars elsewhere, such as on your work computer. So I also show you how to combine all your calendars with your Galaxy S22 phone. After you read this chapter, you'll have no excuse for missing a meeting. (Or, okay, a date.)

TIP

Some calendars use the term *appointments* for *events.* They are the same idea. I use the term *events.*

## Syncing Calendars

Most likely, you already have at least two electronic calendars scattered in different places: a calendar tied to your work computer and a personal calendar. Now you have a third one — the one on your Samsung phone that is synced to your Gmail account.

Bringing together all your electronic calendars to one place, though, is one of the best things about your phone — as long as you're a faithful user of your electronic calendars, that is. To begin this process, you need to provide authorization to the respective places that your calendars are stored in the same way as you authorized access to your email accounts and contacts in Chapters 5 and 6. This authorization is necessary to respect your privacy.

If your phone doesn't have a Calendar icon on the Home screen, open the Calendar app from your App list. This same app works with the calendar that's stored on your phone and any digital calendars that you add.

When you first open this app, you see a calendar in monthly format, as shown in Figure 13-1. I discuss other calendar views in the next section.

FIGURE 13-1:
The monthly calendar display.

When you add an account to your phone, such as your personal or work email account, your Facebook account, or Dropbox, you're asked whether you want to sync your calendar. The default setting for syncing is typically every hour.

Unless you get a warning message that alerts you to a communications problem, your phone now has the latest information on appointments and meeting requests. Your phone continues to sync automatically. It does all this syncing in the background; you may not even notice that changes are going on.

**WARNING**

You could encounter scheduling conflicts if others can create events for you on your digital calendar. Be aware of this possibility. It can be annoying (or worse) to think you have free time, offer it, and then find that someone else took it.

# Setting Calendar Display Preferences

Before you get too far into playing around with your calendar, you'll want to choose how you view it.

If you don't have a lot of events, using the month calendar shown in Figure 13-1 is probably a fine option. On the other hand, if your day is jam-packed with personal and professional events, the daily or weekly schedules might prove more practical. Switching views is easy. For example, just tap the three lines at the top of the calendar, which in this case displays April 2022, to bring up the options. These are shown in Figure 13-2.

If you tap Week, you see the weekly display, as shown in Figure 13-3.

**FIGURE 13-2:**
The Calendar
Display
Options
pop-up menu.

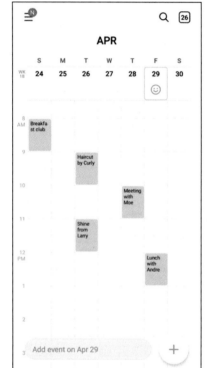

**FIGURE 13-3:**
The weekly
calendar
display.

Or tap the Day button at the top of the calendar to show the daily display, as shown in Figure 13-4.

**FIGURE 13-4:**
The daily
calendar
display.

The following is a representation of the calendar display:

**APR**

**26** Tuesday

7 AM

8

9
Haircut by Curly

10

11
Shoe Shine from Larry

12 PM
Lunch with Chef Ramsey

1

2

3
Meeting with Moe

4

5

6

7

8
Add event on Apr 26          +

9

10

To see what events you have upcoming, regardless of the day they're on, you might prefer List view. Tap the List button at the top of the calendar to see a list of your activities.

TIP

The weekly and daily calendars show only a portion of the day. You can slide the screen to see the time slots earlier and later. You can also pinch and stretch to see more or less of the time slots.

There is also an annual calendar, as seen in Figure 13-5. This is so busy that it does not allow you to see any appointments. It is primarily useful for setting dates out in the future.

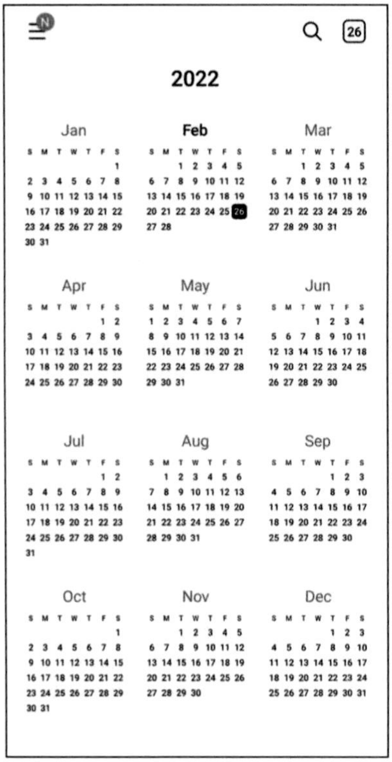

**FIGURE 13-5:**
The annual calendar display.

# Setting Other Display Options

In addition to the default display option, you can set other personal preferences for the calendar on your phone. To get to the settings for the calendar, tap the three lines you used to get to the daily, weekly, or monthly calendars setting options. Scroll down to the bottom, and tap the Manage Calendar button. The screen shown in Figure 13-6 appears.

You have several options. The first is to add a new account. Many people have multiple accounts, including those associated with multiple email accounts. This option makes it convenient to add any other calendars that you haven't already added with your email accounts.

The second option is to remove a calendar. You can just tap the calendar name to make it go away. This is handy if you have too many calendars competing for your attention. You can add the deleted calendar back by tapping its name.

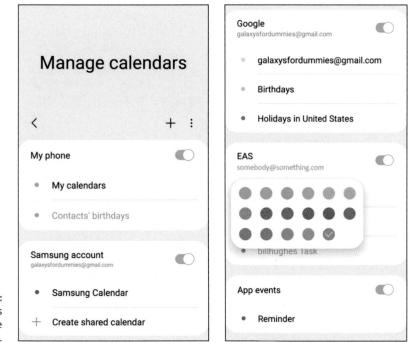

FIGURE 13-6:
The Settings
options for the
Calendar app.

You can also show completed reminders. You may want to be reminded that you took care of something, or maybe you want to forget about it and move on. Your choice. These options allow you to get the right level of detail for your busy life without cluttering it up with too much information.

# Creating an Event on the Right Calendar

An important step in using a calendar when mobile is creating an event. It's even more important to make sure that the event ends up on the right calendar. This section covers the steps to make this happen.

## Creating, editing, and deleting an event

Here's how to create an event — referred to as (well, yeah) an "event" — on your phone. Start from one of the calendar displays shown in Figures 13-1, 13-3, or 13-4. Tap the blue plus sign (+)at the bottom right of the calendar. Doing so brings up the pop-up screen shown in Figure 13-7 (without the keyboard).

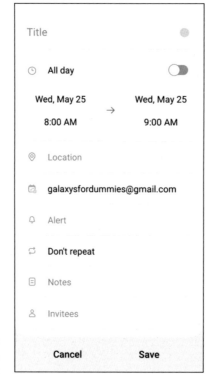

The only required information to get things started is a memorable event name and the Start and End times.

You need the following information at hand when making an event:

» **Calendar:** Decide on which calendar you want to keep this event. Figure 13-7 shows that this event will be stored on the Gmail calendar. This is indicated by showing the amber dot to the right of the title and my Gmail account in the middle of the screen. When you tap the email account, the phone presents you with the other calendars you've synced to your phone; you can store an event on any of them.

» **Title:** Call the event something descriptive so that you can remember what it is without having to open it up.

» **Start and End:** The two entry fields for the start and end times. Select the All Day toggle if the event is a full-day event.

>> **Alert:** This field lets you set how soon before an event your phone will alert you that you have an upcoming event. This makes it less likely that you'll be late to an appointment if you set this alert to accommodate your travel time.

>> **Location:** This setting ties in to the Google Maps app, allowing you to enter an address or a landmark. You can also just enter text.

If you want, you can enter more details on the meeting by adding the following:

>> **Repeat:** This option is useful for recurring events, such as weekly meetings.

>> **Invitees:** This allows you to have a group meeting by entering multiple contacts.

>> **Notes:** Add information that you find useful about that meeting.

>> **Time zone:** If you and everyone who is invited will be in the same time zone, you are set. If you happen to be in another time zone or some of the invitees are in other time zones, you can use this option to ensure that you are not setting a meeting off business hours, or worse, in the middle of the night.

After you fill in the obligatory (and any optional) fields and settings, tap the Save link at the bottom of the page. The event is stored in whichever calendar you selected when you sync.

After you save an event, you can edit it, share it, or delete it by tapping it on any of your calendars. It appears as shown in Figure 13-8. Now you can

>> **Edit:** This brings up the information as entered. Make your changes and tap Save.

>> **Share:** You can send this event as an attachment to anyone as a text.

>> **Delete:** To delete an event, you simply tap the garbage can icon at the bottom of the screen and the event is gone.

TIP

You can also create an event by tapping the calendar itself twice. Tapping it twice brings up a pop-up (refer to Figure 13-7) where you can enter the event details.

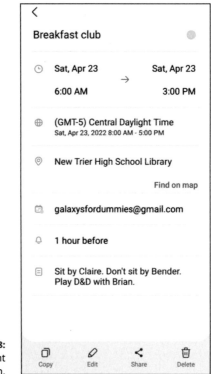

**FIGURE 13-8:**
The Edit Event
screen.

Breakfast club

Sat, Apr 23      Sat, Apr 23
6:00 AM      →      3:00 PM

(GMT-5) Central Daylight Time
Sat, Apr 23, 2022 8:00 AM - 5:00 PM

New Trier High School Library

Find on map

galaxysfordummies@gmail.com

1 hour before

Sit by Claire. Don't sit by Bender.
Play D&D with Brian.

Copy      Edit      Share      Delete

# Keeping events separate and private

When you have multiple calendars stored in one place (in this case, your phone), you might get confused when you want to add a new event. It can be even more confusing when you need to add the real event on one calendar and a placeholder on another.

Suppose your boss is a jerk and to retain your sanity, you need to find a new job. You send your resume to the arch-rival firm, Plan B, which has offices across town. Plan B is interested and wants to interview you at 3 p.m. next Tuesday. All good news.

The problem is that your current boss demands that you track your every move on the company calendaring system. His draconian management style is to berate people if they're not at their desks doing work when they're not at a scheduled meeting. (By the way, I am not making up this scenario.)

You follow my drift. You don't want Snidely Whiplash trudging through your calendar, sniffing out your plans to exit stage left, and making life more miserable if Plan B doesn't work out. Instead, you want to put a reasonable-sounding placeholder on your work calendar, while putting the real event on your personal calendar. You can easily do this from your calendar on your Samsung Galaxy S22. When you're making the event, you simply tell the phone where you want the event stored, making sure to keep each event exactly where it belongs.

The process begins with the Create Event screen. You bring this up by tapping the + sign in the green circle seen at the bottom-right corner of any of the calendars seen in Figures 13-1, 13-3, or 13-4. The information for the real event is shown on the left in Figure 13-9. The fake event is shown on the right. This is the one that is saved to your work email. The colored dot by the work address helps you be aware that this is a different calendar.

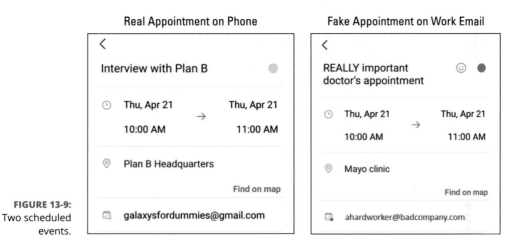

**FIGURE 13-9:** Two scheduled events.

Now, when you look at your calendar on your phone, you see two events at the same time. (Check it out in Figure 13-10.) The Galaxy S22 doesn't mind if you make two simultaneous events.

Under the circumstances, this is what you wanted to create. As long as your boss doesn't see your phone, you're safe — to try to find more fulfilling employment, that is.

**FIGURE 13-10:**
Two events on the same day on your phone calendar.

IN THIS CHAPTER

» **Deciding what you want to use for navigation**

» **Using what's already on your phone**

» **Using maps safely**

# Chapter **14**

# Mapping Out Where You Want to Be

Having a map on your phone is a very handy tool. At the most basic level, you can ask your phone to show you a map for where you plan to go. This is convenient, but only a small part of what you can do.

With the right apps, your Galaxy S22 phone can do the following:

» Automatically find your location on a map.

» Tell others where you are.

» Get directions to where you want to go.

- As you drive, using historical driving times.

- As you drive, using real-time road conditions.

- While you walk, ride a bike, or take public transportation.

» Find new places to go.

» Use the screen on your phone as a viewfinder to identify landmarks as you pan the area (augmented reality).

There are also some mapping apps for the Galaxy S22 for commercial users, such as TruckMap from TruckMap or SmartTruckRoute2 from Smart-Routing, but I don't cover them in this book.

# GPS 101: First Things First

You can't talk smartphone mapping without GPS in the background, which creates a few inherent challenges of which you need to be aware. First off (and obviously), there is a GPS receiver in your phone. That means the following:

>> **Gimme a sec.** Like all GPS receivers, your location-detection system takes a little time to determine your location when you first turn on your phone.

>> **Outdoors is better.** Many common places where you use your phone — primarily, within buildings — have poor GPS coverage.

>> **Nothing is perfect.** Even with good GPS coverage, location and mapping aren't perfected yet. *Augmented reality,* the option that identifies local landmarks on the screen, is even less perfect.

>> **You must be putting me on.** Your GPS receiver must be turned on for it to work. Sure, turning it off saves battery life, but doing so precludes mapping applications from working.

>> **Keep it on the down-low.** Sharing sensitive location information is of grave concern to privacy advocates. The fear is that a stalker or other villain can access your location information in your phone to track your movements. In practice, there are easier ways to accomplish this goal, but controlling who knows your location is still something you should consider, particularly when you have apps that share your location information.

**REMEMBER**

Good cellular coverage has nothing to do with GPS coverage. The GPS receiver in your phone is looking for satellites; cellular coverage is based upon antennas mounted on towers or tall buildings.

**TIP**

Mapping apps are useful, but they also use more battery life and data than many other apps. Be aware of the impact on your data usage and battery life. Leaving mapping apps active is convenient, but it can also be a drain on your battery and your wallet if you don't pay attention to your usage and have the wrong service plan.

# Practically Speaking: Using Maps

The kind of mapping app that's easiest to understand is one that presents a local map when you open the app. Depending on the model of your phone, you may have a mapping apps preloaded, such as Google Maps, Waze, or MapQuest. You can find them on your Home screen and in your Application list.

It's not a large leap for a smartphone to offer directions from your GPS-derived location to somewhere you want to go in the local area. These are standard capabilities found in each of these apps.

REMEMBER

This section describes Google Maps, which is free and may come preinstalled on your phone. If not, you can download it from the Google Play Store. Other mapping apps that may come with your phone, such as Bing Maps or Scout GP Navigation, have similar capabilities, but the details will be a bit different. Or you may want to use other mapping apps. That's all fine.

In addition to the general-purpose mapping apps that come on your phone, hundreds of available mapping apps can help you find a favorite store, navigate waterways, or find your car in a crowded parking lot.

WARNING

As nice as mapping devices are, they're too slow to tell you to stop looking at them and avoid an oncoming car. If you can't control yourself in the car and need to watch the arrow on the map screen move, do yourself a favor and let someone else drive. If no one else is available to drive, be safe and don't use the navigation service on your phone in the car.

The most basic way to use a map is to bring up the Google Maps app. The icon for launching this app is shown here.

The first screen that you see when you tap the Google Maps icon is a street map with your location. Figure 14-1 shows an example of a map when the phone user is in the Seattle area.

Your location is the blue dot at the center of the map — unless you're moving, at which point it becomes a blue arrow. The resolution of the map in the figure starts at about one square mile. You can see other parts of the map by placing a finger on the map and dragging away from the part of the map that you want to see. That brings new sections of the map onto the screen.

TIP

Turn the phone to change how the map is displayed. Depending on what you're looking for, a different orientation might be easier.

Search Box   Voice Commands   Map Menu

Restaurants   Gas   Groceries

Local Services   Directions

Centering Icon

**FIGURE 14-1:**
You start
where
you are.

## Changing map scale

A resolution of one square mile will work, under some circumstances, to help you get oriented in an unfamiliar place. But sometimes it helps to zoom out to get a broader perspective or zoom in to find familiar landmarks, like a body of water or a major highway.

To get more real estate onto the screen, use the pinch motion I discuss in Chapter 2. This shrinks the size of the map and brings in more of the map around where you're pinching. If you need more real estate on the screen, you can keep pinching until you get more and more map. After you have your bearings, you can return to the original resolution by double-tapping the screen.

On the other hand, a scale of one square mile may not be enough. To see more landmarks, use the stretch motion to zoom in. The stretch motion expands the boundaries of the place where you start the screen. Continue stretching and stretching until you get the detail that you want. Figure 14-2 shows a street map both zoomed in and zoomed out. The map on the left is zoomed in in Satellite view. The map on the right is zoomed out in Terrain view.

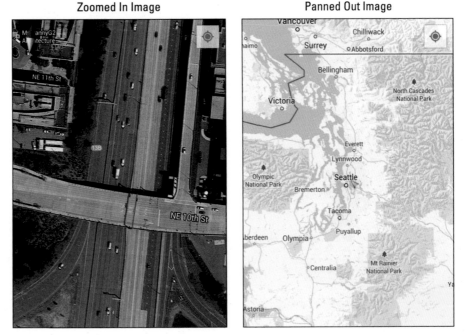

FIGURE 14-2:
A street image
zoomed in and
zoomed out.

The app gives you the choice of Satellite view or Terrain view by tapping the menu button, the circle with the initial of your first name, on the top-right corner of the map. This brings up a pop-up menu similar to the one shown in Figure 14-3.

To enable the Satellite view, first tap Settings. The Settings options are shows in Figure 14-4.

You enable this view by tapping the Satellite toggle switch. You can select a number of options that are useful to you, including transit routes and bicycling paths. I show you some of the other options in the next section in this chapter.

TIP

If you're zooming in and can't find where you are on the map, tap the dot-surrounded-by-a-circle icon (refer to Figure 14-1 for the centering icon). It moves the map so that you're in the center.

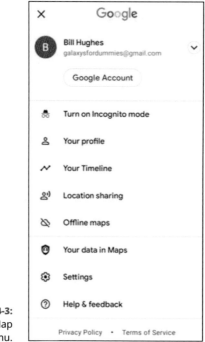

**FIGURE 14-3:**
The Map
menu.

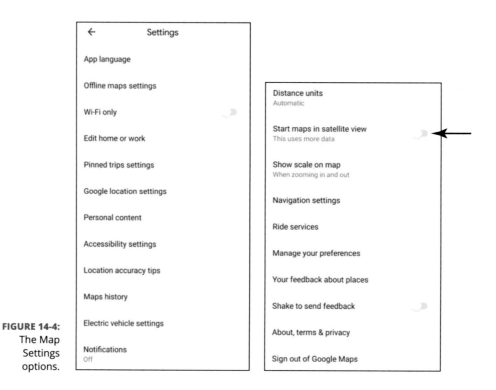

**FIGURE 14-4:**
The Map
Settings
options.

# Finding nearby services

Most searches for services fall into a relatively few categories. Your Maps app is set up to find what you're most likely to seek. By tapping the Local Services icon at the bottom of the page (refer to Figure 14-1), you're offered a quick way to find the services near you, such as restaurants, coffee shops, bars, hotels, attractions, ATMs, and gas stations, as shown in Figure 14-5.

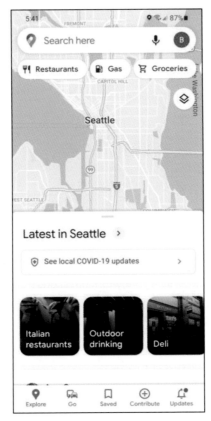

**FIGURE 14-5:**
Tap to find
a service on
the map.

Not only that, it is aware of the time of day. There are different suggestions during breakfast time than in the evening. Just scroll down and tap one of the topical icons, and your phone performs a search of businesses in your immediate area. The results come back as a regular Google search with names, addresses, and distances from your location. An example is shown in Figure 14-6.

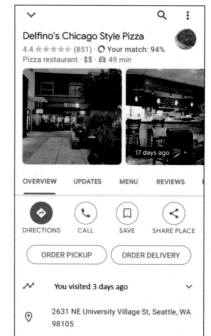

**FIGURE 14-6:**
The results
of a service
selection.

In addition to the location and reviews, the search results include icons and other relevant information:

» **Directions:** Tap the car icon to get turn-by-turn directions from your location to this business.

» **Call:** Tap this to call the business.

» **Save:** Tap the star to set this place as one of your favorites.

» **Website:** Tap this to be taken to the website for this business.

» **Share Place:** Tap the Share Place icon, and you're presented with lots of ways to tell your friends, family, co-workers, and the world about this wonderful place.

» **Order Online:** This option is available if the restaurant owners have enabled it.

» More options, which include

- *Street View:* See the location in Google Street View. As shown at the bottom of Figure 14-6, Street View shows a photo of the street address for the location you entered.

- *Hours:* If this establishment has shared its hours of operation, you can find them here.

- *Menu:* If this establishment has shared its menu, you can find it here.

- *Reviews:* This includes all kinds of information about what people have experienced at this location.

- *More:* Run another Google search on this business to get additional information, such as reviews from other parts of the web.

But let's say you show up, and the line is out the door. It happens. All you need to do is tap the street address, and the local map will appear, as shown in Figure 14-7. This map tells you all the locations around you so you can find another option.

**FIGURE 14-7:** A street map search result.

# Getting and Using Directions

You probably want to get directions from your map app. I know I do. You can get directions in a number of ways, including:

>> Tap the Search text box and enter the name or address of your location — for example, Seattle Space Needle or 742 Evergreen Terrace, Springfield, IL.

>> Tap the Explore icon (refer to Figure 14-1), and select your location.

Any of these methods lead you to the map showing your location.

TIP

Sometimes the app gives audio instructions. The Google Maps app has a speaker icon that appears on the map that allows you to turn audio instructions on or off. It might seem intuitive to expect that when you search for a specific attraction (such as the Seattle Space Needle), you get only the Seattle Space Needle. Such a result, however, is sometimes too simple. A given search may have multiple results, such as for chain stores. Google Maps gives you several choices, and you may need to choose your favorite. Typically, the app will give you the closest option.

To get directions, tap the Directions icon. This brings up the pop-up screen shown in Figure 14-8.

This gives you the options of

>> **Driving:** Turn-by-turn directions as you drive from where you are to the destination.

>> **Public Transportation:** This option tells you how to get to your destination by taking public transportation using published schedules.

>> **Taxi or Ride Service:** This option gives the time if you were to hop a taxi, Uber, or Lyft.

>> **Cycling:** This option is for the cyclists among us; it includes bike trails in addition to city streets.

>> **Walking Navigation:** Turn-by-turn directions as you walk to your destination.

**FIGURE 14-8:**
Your direction options, from original location to the target.

## DRIVING SAFELY

Using your phone while driving is a bad idea. Texting while driving is illegal in most states, but studies have shown that laws don't affect behavior that much. It's just so darn tempting to sneak a peek at your phone. The grim reality: According to the National Highway Transportation Administration, texting while driving resulted in just under 600 deaths in the United States in 2016. When you include all the sources of distracted driving (eating, drinking, adjusting the radio, and so on), 3,450 people were killed.

To make matters worse, distracted driving is disproportionately problematic for younger drivers. You can't do much to stop your teen from checking out their hair in the rearview mirror while they're driving, but you can install an app on their phone that will prevent them from texting while the car is in motion. The simplest apps track speed — if the phone is going above some preset limit, it shuts off the phone's communication. More sophisticated apps give you more options, including the ability to place calls to 911 or a list of exceptions, a code that you can enter if you're the passenger and not the driver, and more.

If you have trouble setting down your phone, you may even want to install one of these apps for yourself *and* for your teen.

For each of these options, you can use the options at the bottom of the screen to

» **Get Directions:** Sequential directions, as shown in Figure 14-9, but without telling you when to turn.

» **Navigate:** Rather than show you a map, this option puts you in a navigation app that monitors where you are as you travel and tells you what to do next.

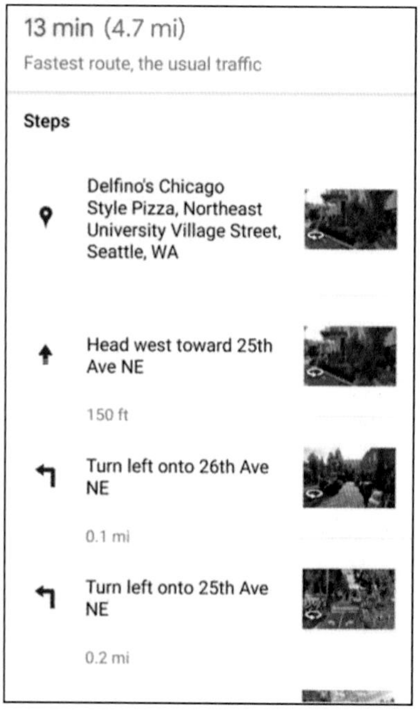

FIGURE 14-9:<br>Step-by-step<br>directions to<br>the target.

13 min  (4.7 mi)
Fastest route, the usual traffic

**Steps**

Delfino's Chicago Style Pizza, Northeast University Village Street, Seattle, WA

Head west toward 25th Ave NE

150 ft

Turn left onto 26th Ave NE

0.1 mi

Turn left onto 25th Ave NE

0.2 mi

# Chapter **15**

# Paying with Samsung Pay

S omeday in the future, you will be able to leave your wallet at home and just use your smartphone for all the stuff that you carry with you today. We are not there yet, but Samsung has implemented some technology that moves us toward that day.

You may reasonably ask why not carrying your wallet is such a big deal. Fair enough. The two main reasons are that it is more convenient and more secure. Samsung Pay takes us a long way to this ultimate goal. The reality is you will need to continue to carry your wallet with credit cards and cash for now.

For now, you'll need to decide for yourself whether the convenience that Samsung Pay does offer is worth the effort associated with setting it up. Samsung has gone to great lengths to make transactions more secure than simply using a credit card. Like so many of these things, the security is only as good as how the individual implements it.

With that awareness, some things about Samsung Pay are exceptionally cool. If you like to show off new technology to people, this is a good opportunity. In particular, you can one-up your smug friends who use Apple Pay with their iPhones. That capability alone is worth the price of admission for some of us.

# How Mobile Payment Works

It helps to explain how mobile payments work to see how Samsung Pay helps you. The idea is that you walk up to the checkout stand at a retailer. You place your finger on the screen, place your phone near the card reader, and instantly your transaction is done. The receipt prints, and you take your purchase and go on your merry way. The charge shows up on your credit card.

Besides a fingerprint, you can enter a four-digit PIN. You choose the security option that is most convenient for you, and Samsung makes sure that everything flows securely. All the authorizations were set up when you set up the credit card with Samsung Pay.

Really. That's all there is.

It sounds so simple. The interesting thing is that this capability has been around in one form or another for close to 30 years. Back in 1997, Mobil gas stations implemented contactless payment with their Speedpass. Users on the system would tap their Speedpass fobs that they would keep on the key ring on the reader installed on the gas pump.

Their credit card would be charged for the gas they pumped. Initially, the tap was all that it took to pay for the gas. Later, users needed to add their ZIP code to enhance security. In spite of Mobil's best efforts, the system had only moderate success. There are a range of reasons, but a major concern was that people were already hesitant to even carry yet another thing. It is the inclusion of some new technology within smartphones that makes it so much more convenient.

Visa and MasterCard also tried contactless systems with limited success. One of their major challenges is that not enough credit card readers support the contactless payment system. After years of trying, fewer than 10 percent of all credit card readers as recently as 2015 have this capability. Those readers that do support contactless payment have the logo shown in Figure 15-1.

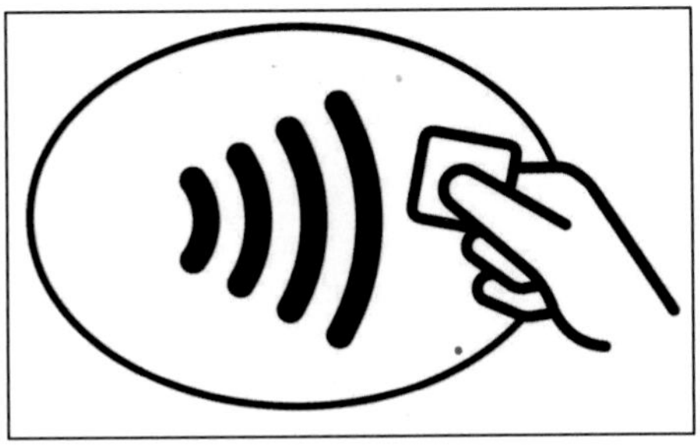

**FIGURE 15-1:**
The contactless
payment logo.

The contactless payment approach got a significant boost when Apple introduced the Apple Pay system on the iPhone in 2014. Like some other technologies that have struggled to find widespread acceptance, the implementation by Apple produced enough momentum to overcome resistance to its widespread use.

Then came Samsung Pay where Samsung leapfrogged Apple. Samsung Pay works similarly to Apple Pay with contactless payment devices. It also works with the vast majority of standard credit card readers. Take that, Apple! Samsung Pay took off.

At about this time, the credit card industry embraced the use of the embedded chip. This meant that most retailers needed to upgrade their readers. In addition to supporting the chip reader, the new readers supported contactless payment. Samsung Pay along with your Galaxy S22 have the electronics built in to work with the contactless system! Using the contactless system lets you avoid the hassle of finding the slot and orienting the card in the right direction.

**TECHNICAL STUFF**

The chip that is now on most credit cards is called an EMV chip. It quickly and securely sends some double-secret codes that verify that this card is the original card sent to the account owner and not created by a thief in their basement. The use of Samsung Pay provides the same level of confidence to the credit card company, so it's just as happy that you use Samsung Pay as the EMV chip. Everyone wins, including the thief who now must go find gainful employment!

Probably the biggest drawback in using this technology these days is that other customers in line freak out the first time they see this slick solution. They will want to ask you more about it, which will end up taking more time than if you were paying by writing a check and having to provide two forms of identification!

# Getting Started with Samsung Pay

The first step in the process is to make sure that you have Samsung Pay on your Galaxy S22. As cool as this app is, there are many options to this technology, and your carrier may have preferred to not have it preloaded. No problem. Chapter 8 explains how to download apps from the Play Store. The Samsung Pay logo is seen in Figure 15-2.

**FIGURE 15-2:**
The Samsung Pay and the Google Pay logos.

**TIP**

It is easy to confuse Samsung Pay and Google Pay. These are two different applications. Google Pay is nice, but it does not have all the cool features of Samsung Pay.

The app page description is shown in Figure 15-3. When you tap Install, you get the image on the right.

**FIGURE 15-3:**
The Samsung Pay app Play Store images before and after installation.

The Samsung Pay app works a little different than most other apps. The first time you open Samsung Pay, you will be given information on how to use it and be asked to put in your credit card information (in a very convenient way, by the way) and asked all kinds of permissions and agreements.

Once you have given all this information, Samsung Pay waits patiently at the bottom of your home pages, ready to meet your payment needs with nothing more than a quick swipe from the bottom of the screen. Most people would simply not use this app if they had to go digging through their screens to find the app. This way, you do not have to search to find the app and wait for it to come up.

Figure 15-4 shows the Home screen with the Samsung Pay launch button sitting at the bottom, ready to appear with a quick flick.

**FIGURE 15-4:**
The Samsung
Pay Quick
Launch button.

Samsung Pay
Quick Launch
button

The launch button is also there on the Lock screen, so you don't even need to unlock your phone because you'll be using the exact same security steps before you can use your credit card.

# Setting Up Samsung Pay

When you open the Samsung Pay app, you're greeted with a series of pages before you get to the Home screen seen in Figure 15-5. These pages include marketing introductions (which you don't need because you read this chapter), permissions and agreements (which you should give if you want to move forward), and some pages that verify that Samsung Pay can work on your phone.

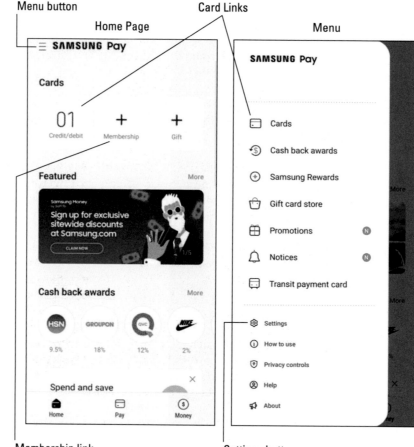

**FIGURE 15-5:**
The Samsung
Pay Home
page.

The app wants to make sure that your phone has the right parts (your Galaxy S22 does, but that is not the case for every Android phone) and that you are in one of the 29 countries where Samsung Pay is accepted. You're set if you're in the United States, Canada, China, or Kazakhstan (very nice!). You're out of luck if you're in, say, Yemen. Being in these countries is important for this app because each country has its own set of laws for payments, and Samsung may not have all the arrangements in place if you are not in one of these countries. New countries will be added as quickly as possible.

TIP

Read the agreements. In all likelihood, they are the same kind of agreements that exist in the fine print in your existing credit card and every time you sign your name to a credit card charge slip that affirms you won't pull any shenanigans. That said, I am your humble author, not your legal advisor, and you should feel comfortable with these agreements.

The next part of the initial setup process is to step through how you want to set up security. You have two security options when you are making a transaction: using your fingerprint or entering a PIN.

There are a few scenarios where using a Samsung Pay PIN rather than a fingerprint might make sense. In most cases, using your fingerprint is exceptionally convenient. If you haven't set these up yet, the Samsung Pay app will walk you through the steps to do so. It's quick and easy.

The next step is to enter your credit card information. When you tap the link that says Cards on the home page or the menu, you're taken to a screen that shows you the images of the credit cards you've added. Initially, the screen will be empty. Figure 15-6 shows what it looks like after you have entered your first card.

To add a new card, tap the blue circle in the lower-right corner with a silhouette of a credit card and a plus sign. You get a pop-up with some choices. Pick the Add Credit/Debit Card option to start this process. The left image in Figure 15-7 is when the screen first comes up and the right image in Figure 15-7 is when you have the desired credit card in the viewfinder.

The app then interprets the information on the front of your credit card and populates as many screens as possible with your credit card information. It will ask you to fill out the form seen in Figure 15-8 if it can't figure out the information on your card or the information is on the reverse side of the card.

**FIGURE 15-6:**
The Samsung
Pay Card page.

Icon to add a card

Before taking an image of your card        After taking an image of your card

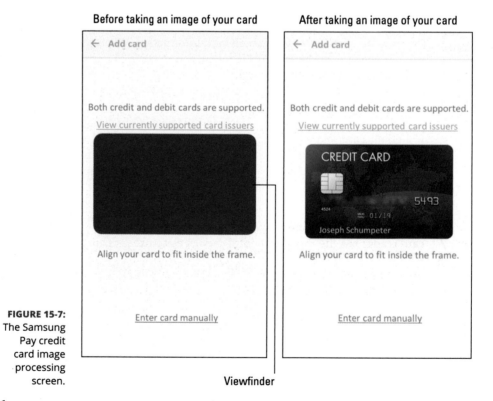

**FIGURE 15-7:**
The Samsung
Pay credit
card image
processing
screen.

Viewfinder

Not every company that offers credit cards is signed up with Samsung Pay. You can check the Samsung website at www.samsung.com/us/samsung-pay/compatible-cards/#bank or tap the link in Figure 15-8 to check before you proceed. Otherwise, just go ahead and see whether things go through.

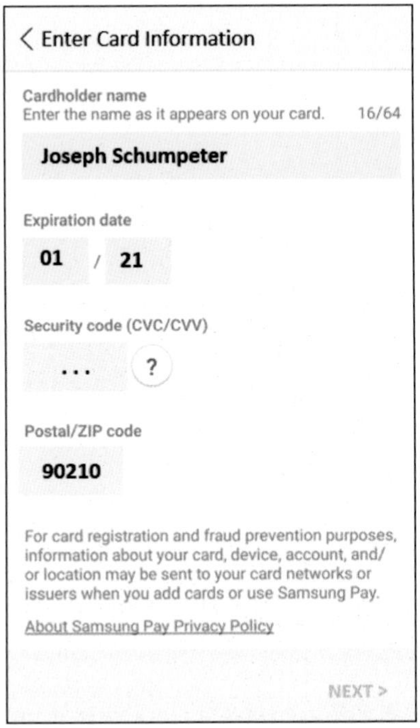

< Enter Card Information

Cardholder name
Enter the name as it appears on your card.    16/64

**Joseph Schumpeter**

Expiration date

**01**   /   **21**

Security code (CVC/CVV)

**. . .**    ?

Postal/ZIP code

**90210**

For card registration and fraud prevention purposes, information about your card, device, account, and/ or location may be sent to your card networks or issuers when you add cards or use Samsung Pay.

About Samsung Pay Privacy Policy

NEXT >

**FIGURE 15-8:**
The Populated credit card data fields.

Tap Next, and Samsung Pay will seek to confirm things with your credit card company. This does not cost you anything. It just wants to make sure that when you do make a charge, everything will flow smoothly. One of the things the company will want to verify is that you are authorized to use that credit card. This means that either you need to be the primary cardholder or you need to coordinate with that person.

Keep in mind that scanning your fingerprint and tapping your phone on a credit card reader has the same legal implications as actually signing a credit slip.

When the credit card company verifies that everything is on the up and up, you get an acknowledgement like the one shown in Figure 15-9.

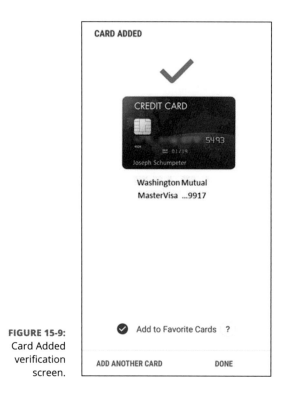

**FIGURE 15-9:**
Card Added
verification
screen.

# Using Samsung Pay

First, pick something to buy at a store. Have a clerk ring it up and tell them you will pay with a credit card. Swipe the screen upward. You see the screen seen in Figure 15-10.

Enter your fingerprint by pressing the icon and you see a screen similar to Figure 15-11.

As seen in Figure 15-11, you should either hold your phone against where the credit card reader would read the magnetic stripe or where there is the contact-less payment system logo if it is available.

**FIGURE 15-10:**
Samsung Pay payment screen before reading your fingerprint.

The semicircle above your signature will start counting down, and your phone will vibrate to let you know that it is transmitting. You will hear a beep if it all goes through. If it fails for some reason, you can try again simply by reentering your security option.

TIP

This process works with the vast majority of credit card readers. This does not work well, however, with credit card readers where you insert your card into the machine and pull it out quickly. This type of reader is mostly used on gasoline pumps. More and more gasoline pumps have a contactless reader, but there are many generations of pumps out there, so your mileage may vary. Sorry.

# Managing Samsung Pay

As long as you pay your credit card bills and keep your credit card account at the bank, this app is relatively self-sufficient. Still, you will need to access the app settings from time to time. You get there by tapping the Samsung Pay logo on the app screen. This will bring up the screen shown in Figure 15-5.

To add a new credit or debit card, tap the Add link and go through the process described in the section "Setting Up Samsung Pay," earlier in this chapter.

Adding a second card is easy, and the images will stack up when above each other on this page. The quick launch link brings up whatever card is on the top. If you want to change to use a different credit card, you can flip through the options. In addition to credit and debit cards, Samsung Pay is happy to help you with membership cards and loyalty cards. Don't be shy. Give it a try and see if you like the convenience. My bet is that you will.

You tap the three dots to get a pop-up for the settings of Samsung Pay and select Settings. The Settings page is shown in Figure 15-12.

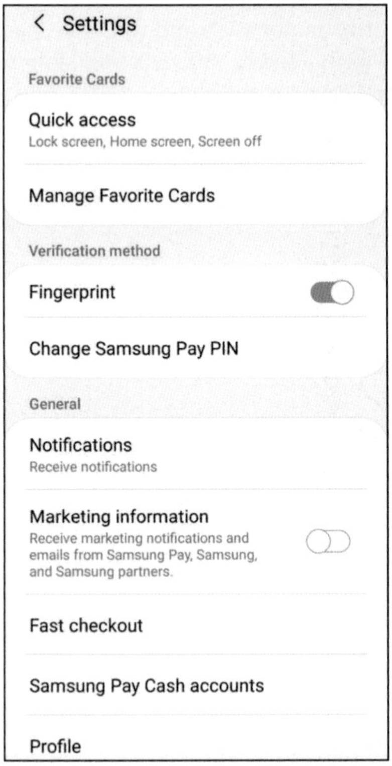

**FIGURE 15-12:**
Samsung
Pay Settings
screen.

If you want to change the settings for or delete a credit card, all you have to do is tap on the Manage Favorite Cards link. All the information associated with that card will appear in the screen.

Enjoy your spending!

# Adding Loyalty Cards

Let's say that you are at Macy's with your child shortly before Christmas. Junior tells Santa that he wants a red firetruck, and Santa assures him that he'll get one. You freak because you know Macy's and every other toy store you know is sold out. It's the hot new toy this season.

"Tut-tut," Santa reassures you — they're available at Shoenfeld's on Lexington Avenue. The first crisis is averted. However, while you brought your membership card for Macy's, you left your membership card for Shoenfeld's at home.

That's when you realize that this is not 1947 and that you scanned your Schoenfeld's membership card into your Galaxy S22. It's available with all your other membership cards. The second crisis is averted.

To add a membership card, start by tapping the Membership link (refer to Figure 15-5). This brings up the screen shown in Figure 15-13.

Listed membership cards

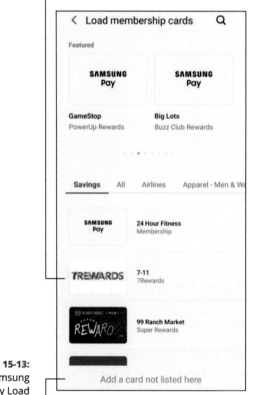

**FIGURE 15-13:** The Samsung Pay Load Membership Cards screen. Unlisted membership cards

This list of membership cards that the app knows is long. Let's say Crocs are your thing. You're in luck! The Crocs Club is one of the options shown in Figure 15-14.

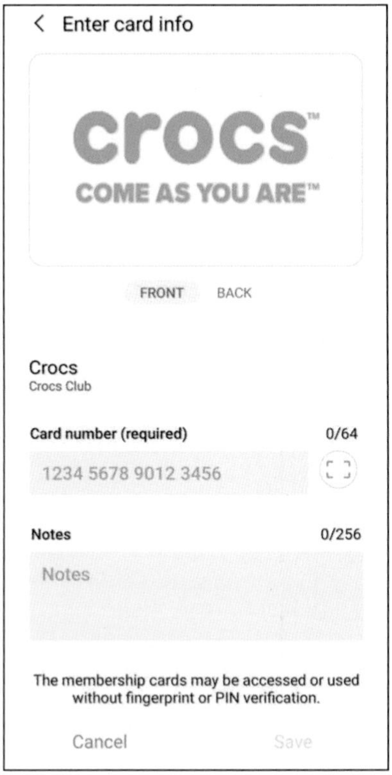

**FIGURE 15-14:**
The Crocs Club membership card in Samsung Pay.

However, it seems as if Schoenfeld's is not among the list of retailers and other organizations on the list. Adding a card that isn't already on the list does not require a miracle anywhere near 34th Street. Just tap the link that says Add a Card Not Listed Here, and you see the screen in Figure 15-15.

Take photos of the front and back of the card, enter some information (like the name of the card and your membership number), and you'll never again need to worry about carrying that stack of membership cards you only use occasionally.

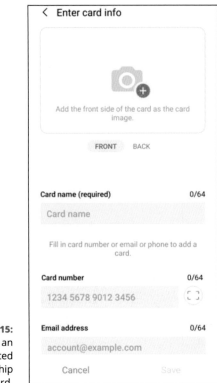

**FIGURE 15-15:**
Adding an
unlisted
membership
card.

# 6

# The Part of Tens

**IN THIS PART . . .**

Get the most out of your phone.

Protect yourself if you lose your phone.

Discover what features will be coming in future versions.

# Chapter **16**

# Ten Ways to Make Your Phone Totally Yours

A smartphone is a very personal device. From the moment you take it out of the box and strip off the packaging, you begin to make it yours. By the end of the first day, even though millions of your type of phone may have been sold, there's no other phone just like yours.

This is the case because of the phone calls you make and because of all the options you can set on the phone and all the information you can share over the web. Your contacts, music files, downloaded videos, texts, and favorites make your phone a unique representation of who you are and what's important to you.

Even with all this "you" on your phone, this chapter covers ten or so ways to further customize your phone beyond what this book covers. Some of these suggestions involve accessories. Others involve settings and configurations. All are worth considering.

When it comes to accessories, your carrier's retail store has many good options. And oodles, heaps, loads, and tons of options for your phone are available outside the carrier's retail store. There is no need to limit yourself.

# Using a Bluetooth Speaker

In just a matter of a couple of years, Bluetooth speakers have gone from an interesting (and expensive) option to mainstream. Prices have come down quickly, and the variety of designs for the speakers has grown dramatically. In all cases, you get the benefit of being able to play your music on your phone without the inconvenience of having headphones. You also get a speakerphone, although the quality of the audio can vary dramatically from Bluetooth speaker to Bluetooth speaker.

This accessory is best purchased in a brick-and-mortar store where you can listen to a variety of choices. Although syncing with a number of speaker choices can be tedious, it's the only way you'll know if a particular Bluetooth speaker meets your needs.

Although sound quality is essential, the folks in the industrial-design department have been having a lot of fun coming up with ways for the Bluetooth speaker to look. Figure 16-1 shows a range of options for the most popular form factors.

Rectangular form factor

Sound Bar form factor

Can form factor

**FIGURE 16-1:** Popular Bluetooth speaker form factors.

*Left: Passakorn Shinark/Shutterstock; Center: ABC vector/Shutterstock; Right: PROFFIPhoto/Shutterstock*

The most popular options are basically a rectangle, the can style, and the sound bar. Each company has its own twist, but these options represent what you can see in the high-volume stores.

Fortunately, you live in a world where designers can get creative. The House of Marley, for example, uses unconventional design and recycled material to stand out in the market. Infinity Orb makes a levitating speaker. More than a few companies make Bluetooth speakers that look like the Death Star from *Star Wars.*

There is no shortage of imagination when it comes to Bluetooth. Be sure to consider all your options. Before you settle on the mainstream options at the local mall, do some research off the beaten path. You'll be amazed at the range of options.

# Cruising in the Car

You may have gotten the idea that I am concerned about your safety when you're using your phone. True, but I'm also concerned with *my* safety when you're using your phone. I would like you to have a Bluetooth speaker in your car when driving in my neighborhood, if you please.

Most new cars have a built-in Bluetooth system within the car stereo that connects to a microphone somewhere on the dash as well as to your car speakers. It's smart enough to sense an incoming call and mutes your music in response.

If you don't have such a setup, there are lots of good options for car speakers. Figure 16-2 shows an example.

**FIGURE 16-2:**
Bluetooth car speaker.

A closely related accessory is a car mount. A *car mount* will hold your phone in a place you can easily observe as you drive. The mount attaches to an air vent, the dashboard, or the windshield.

You could always put your phone on the seat next to you, but that's for amateurs. Instead, offer your phone a place of honor with a car docking station. It makes accessing your phone while you drive safer and easier. Some see getting a car mount as a luxury, but my view is that if you need to upgrade your navigation software, you need a vehicle mount.

Be sure to put away the docking station when you park. Even an empty docking station is a lure for a thief.

Figure 16-3 shows a typical vehicle mount. There are a great variety of options. Some mounts grip the phone, while other involve magnets. Some mounts use suction cups to hold onto the inside of your windshield, some clip on air vents, and still others you can screw into your dashboard. Many allow you to plug in a car charger, while others offer wireless charging. They cost $30 to $50. You can get these at your carrier's store, at Best Buy, or at Amazon.

**FIGURE 16-3:**
A standard vehicle mount.

*Roman Arbuzov/Shutterstock*

More economical vehicle mounts are available that do not include a wireless charging pad. Figure 16-4 shows an example from SlipGrip that costs closer to $30.

This version is special because it is specifically designed to work with larger protective cases. It can be a hassle to have to remove the protective case from your phone to insert it in a vehicle mount. In addition to a car mount, SlipGrip also offers a mount for your bicycle.

When shopping for a car mount or a bike mount, be sure to consider the extra girth associated with a case. If you don't use a case (which is a mistake), you just need to be sure that the mount is compatible with a Galaxy S22. If you have a case on your phone (which is anything but a mistake), be sure the mount will work with the Galaxy S22 *and* the case.

**FIGURE 16-4:**
A SlipGrip
vehicle mount.

*Photograph courtesy of SlipGrip*

# Considering Wireless Charging Mats

Those industrial designers have certainly gone crazy with Bluetooth speakers. They have also been having a field day on wireless charging options.

Most wireless charging mats are designed with the idea that you can place your phone on the mat, and technology will do the rest. The wireless charging pads, shown in Figure 16-5, go for between $35 and $50. These include Fast Charging capabilities, which will charge your phone from stone-cold dead to 100 percent in just a few hours.

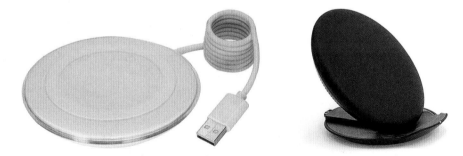

**FIGURE 16-5:**
Wireless
charging pads.

*Left: suthon thotham/Shutterstock; Right: dmitriynesvit/Shutterstock*

If these are too conventional for you, more options are out there as well. Figure 16-6 shows two of the more interesting options: the BOOST↑CHARGE Wireless Charging Stand + Speaker from Belkin (left) and a combination wireless charging stand and LED lamp (right).

BOOST↑CHARGE™ Wireless
Charging Stand + Speaker

Combination Wireless Charging
Stand and LED Lamp

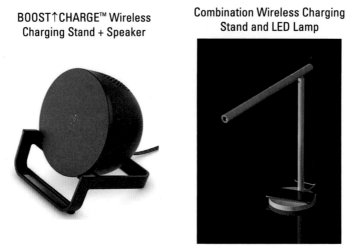

Left: Photograph courtesy of Belkin; Right: Jane Khomi/Getty Images

**FIGURE 16-6:** More exotic wireless charging pads.

Look for more options going forward!

# Making a Statement with Wraps

Not the sandwich kind of wraps. I'm talking about the kind of wrap that lets you customize the non-glass portion of the S22.

The Samsung Galaxy S22 is very attractive, but you can spruce it up with an adhesive covering called a wrap from Skinit at www.skinit.com. These designs can express more of what is important to you. As a side benefit, they can protect your phone from minor scratches.

Figure 16-7 shows some design options for a skin. It comes with cut-outs for speakers, plugs, microphones, and cameras specifically for the Galaxy S22. Putting the skin on is similar to putting on a decal, although it has a little give in the material to make positioning easier. The skin material and adhesive are super high-tech and have enough give to allow klutzes like me (who struggle with placing regular decals) to fit the nooks and crannies of the phone like an expert.

**FIGURE 16-7:**
Some sample
wrap designs
from Skinit.

LUCKY HEART
PATTERN

LEOPARD
PRINT

NEON
CHECKERED

GREEN STREET
CAMO

*Photograph courtesy of Skinit*

If you're not crazy about Skinit's designs, you can make your own with images of your own choosing. Just be sure that you have the rights to use the images!

# You Look Wonderful: Custom Screen Images

In addition to the shortcuts on your extended Home screen, you can also customize the images that are behind the screen. You can change the background to one of three options:

>> Choose a neutral background image (similar to the backgrounds on many PCs) from the Wallpaper Gallery. Figure 16-8 shows a selection of background images that are standard on your phone.

>> Any picture from your Gallery can be virtually stretched across the seven screens of your home page. (Read more about the Gallery in Chapter 9.)

>> Opt for a theme that changes the background and fonts.

The pictures from your Gallery and the Wallpaper Gallery are images that you can set by pressing and holding the Home screen. In a moment, you see icons appear. Select Wallpapers and pick the image you want. Many are free; others you can download for less than a dollar.

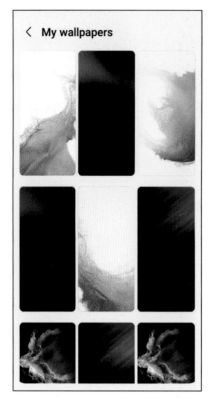

**FIGURE 16-8:**
Standard
wallpaper
sample
images.

The themes take it to a different level. In addition to the background, the phone will change the fonts and many of the basic icons. You can select different themes, as shown in Figure 16-9, that follow different color and font combinations. There are lots of options.

A few samples also come on your phone. You may be a pink kind of person or a space nut. If these options aren't to your taste, tap the Store link at the top of the page to see a range of options. If you do not see one you like today, be sure to check back as developers are adding new themes all the time. Do yourself a favor, and see if any of these strike a chord with you.

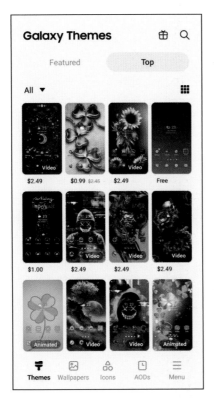

FIGURE 16-9:
Different
theme options.

# Empowering Power Savings

Imagine that you're flying on a short hop from Smallville to Littleton. You're about to take off when air traffic control tells the pilot that the plane needs to wait on the tarmac. Unfortunately, you're a little low on battery and, because it's a flight on a puddle-jumper, there is no charging port. The pilot says it's okay to use your mobile phone.

You don't know how long this will take, so you pull out your phone, go to Settings, tap Battery and Device Care, and tap Battery. From here, you tap the toggle button to put it in Power Savings Mode. This slows down a lot of cool features. For example, it dims the screen, slows the processors and mobile data speeds, and kills the vibration mode. This extends the battery life with minor sacrifices to what you can do with your phone. The apps all run, but perhaps not as fast. If you want to see what you've enabled or you want to disable one of these options, tap the Power Savings Mode link. This brings you to the left screen in Figure 16-10.

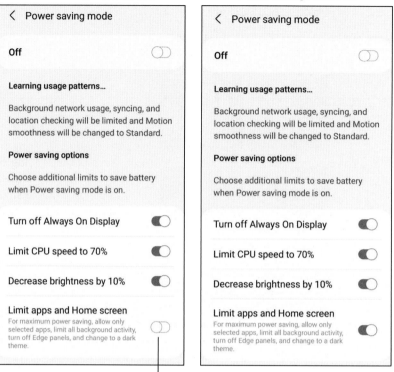

Power Saving Mode Options          Power Savings Mode On

**FIGURE 16-10:**
Power Savings
mode and the
pop-up.

Super Power Savings Mode

From here, you can override one of the three options by tapping the toggle switch. Then you tap the Off toggle switch to On and the phone implements the other power savings options. Hopefully this gets you through the delay. Then the pilot says that your plane has a flat tire, the maintenance crew went home, and you could be sitting there for hours. This time you go back into Settings, tap Battery and Device Care, tap Battery, tap Power Saving Mode, and tap the toggle that says Limit Apps and Home screen. This puts your phone in Super Power Savings mode when you tap Off to make it turn to On.

Your beautiful, powerful smartphone has instantly turned itself into a dumb feature phone. You won't be able to watch a full-length feature film. However, you now have enough battery to make and take phone calls and text messages for hours. This is not an exact science and your mileage may vary, but switching to this mode can multiply the time your smartphone can operate as a phone by eight to ten times. If you aren't ready to return to the 2000s, you can use the regular power savings mode and hope the intelligence of your S22 will be enough to keep your phone alive.

# Controlling Your Home Electronics

Traditional home appliances are getting smarter, offering more capabilities for better performance. One problem they have is that adding rows of control buttons (so that you can control those new capabilities) complicates manufacture. So, the next generation of appliances has begun to add a small LCD screen to the appliance. This can look slick, but it costs a lot and is prone to breakage. Also, the fancy capabilities involve pushing a confusing combination of buttons with cryptic messages displayed on a tiny screen.

The latest idea is to omit the screen altogether and give you control of the settings through an app on your smartphone. Your appliance retains the very basic buttons but allows you to use the fancy capabilities by setting them through your phone. You just download the free app made by the manufacturer from the Google Play Store, and you have your beautiful and logical user interface to control your new product — no strings attached.

For example, your new oven will allow you to turn it on and set the temperature without using a smartphone. However, if you want it to start preheating at 5:30 p.m. so that you can put the casserole in right when you walk in the door, you can set that up through your phone.

# Wearing Wearables

*Wearables* are a class of mobile accessories that have been getting a lot of attention. For the most part, wearables are connected via Bluetooth and typically worn as a watch, although you can wear some sensors in athletic shoes. Samsung offers its wearables under the brand name Samsung Galaxy Watch (see Figure 16-11).

**FIGURE 16-11:**
Samsung
Galaxy Watch
wearables.

In addition to telling the time, these wearables give you notifications, weather, and texts and track some relevant information, such as the amount of exercise you've done. These images look like conventional, but stylish analog watches. It shows you texts, a phone dialing pad, your altitude, and the useful information that would otherwise be shown on your phone. How cool is that?!

We can take it one step further. For example, Neuvena has a product called Xen (see Figure 16-12) that is designed to offer you relaxation and enhanced sleep with its uniquely designed headphones and mobile app. You use the mobile app to adjust the sensations to meet your preferences for relaxation.

This is just another example of how manufacturers are leveraging the capabilities of the smartphone to offer new capabilities far beyond making a call and sending a text.

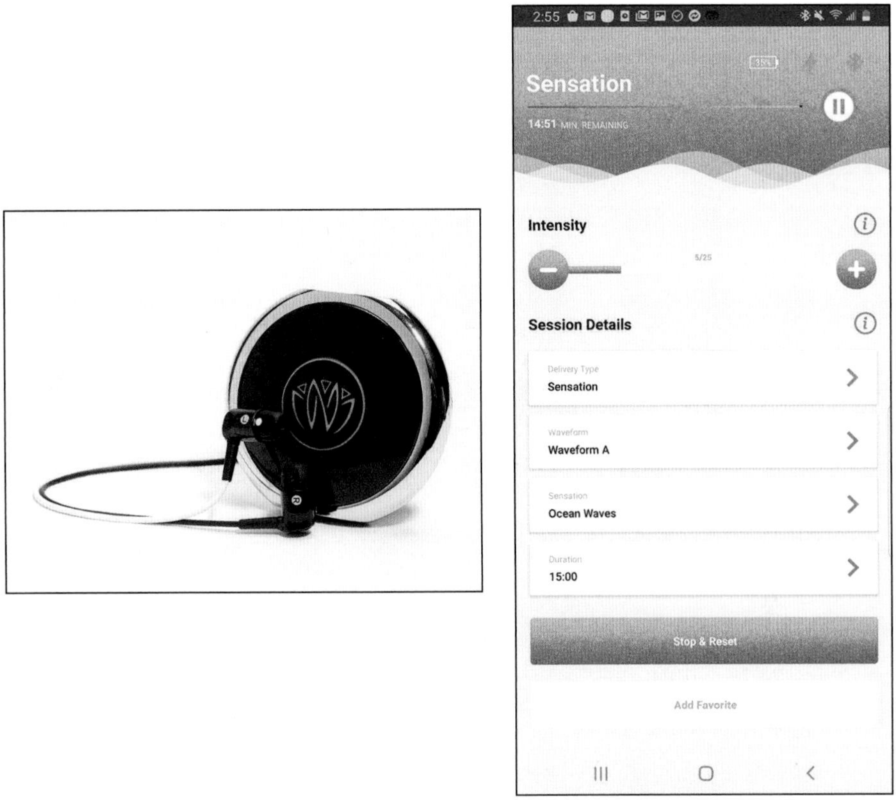


**FIGURE 16-12:** Neuvana Xen wearable.

*Left: Photograph courtesy of Neuvana*

# Using Your Phone as a PC

Your phone has all the computing power of a PC. Wouldn't it be nice if you didn't have to carry a laptop? Enter the Samsung Dex (see Figure 16-13).

The Dex Docking Cable may look like a charging cable, but it's so much more. When you connect the little connector (USB-C) to your phone, you can connect the other end into a PC monitor or a TV. What's on your beautiful small screen now appears on your beautiful big screen! If you connect a Bluetooth keyboard and mouse, it will almost seem like you have a regular PC. This is much smaller and more convenient than lugging around even the lightest laptop.

**FIGURE 16-13:**
Samsung
Dex Cable.

# Creating Your Own AR Emoji in the AR Zone

I've saved the best for last: You can take a self-portrait and make your own emoji that becomes an integral part of your Galaxy S22. You may be used to sending smiley faces, frown faces, and winks. Now you can send *you*.

Let me show you what I mean. Let's say that you're President Teddy Roosevelt, and you're tired after a successful day of assaulting San Juan Hill, busting trusts, and rescuing orphan bear cubs. You sit down with your new Galaxy S22 smartphone, and decide to send a text to your pal Thomas Jefferson. (They were pals, right? They posed together for Mount Rushmore.) Instead of sending the conventional smiley face, you decide to send an AR Emoji. Figure 16-14 shows what you look like in real life and your smiling emoji.

When you create an emoji of yourself, this app creates a series of animations, called GIFs, that provide animation of your emoji showing a variety of emotions. These are stored in the Gallery app, and you can send them to anyone you like.

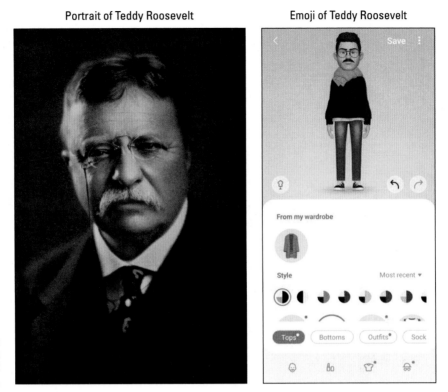

Portrait of Teddy Roosevelt          Emoji of Teddy Roosevelt

**FIGURE 16-14:**
The real image
and the emoji
image.

So, let's say you're happy because you just got word that they completed the Panama Canal. You send a happy emoji. You try, but fail, to institute a federal income tax, so you send a sad emoji. When you're campaigning in Milwaukee, saloon-keeper John Schrank shoots you in the chest, but you keep on talking another 90 minutes. When you're done speechifying, you're quite cross with Mr. Shrank, so you send him an angry emoji. These emojis are shown in Figure 16-15.

If you find your emoji to not quite capture you, you do have the ability to modify the image. You can add or take away hair, as well as adjust the color and style. You can also add glasses (although for some reason they don't have the pince-nez glasses Teddy Roosevelt liked to wear). You can also change attire as needed.

When you're an emoji in your phone, you and your phone are in it for the long haul!

Happy Emoji                    Angry Emoji

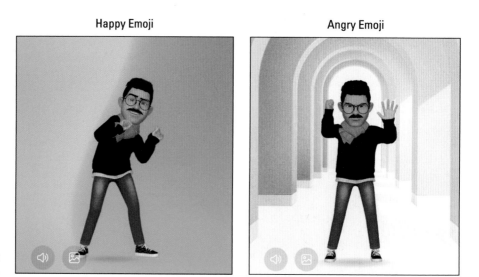

**FIGURE 16-15:**
Emoji options.

> » **Avoiding losing your phone in the first place**
>
> » **Protecting yourself if you do lose your phone**

Chapter **17**

# Ten (or So) Ways to Make Your Phone Secure

**B**ack in the "old" days, it sure was frustrating to have your regular-feature phone lost or stolen. You would lose all your contacts, call history, and texts. Even if you backed up all your contacts, you would have to reenter them in your new phone. What a hassle.

**WARNING**

The good news is that your smartphone saves all your contacts on your email accounts. The bad news is that, unless you take some steps outlined in this chapter, evildoers could conceivably drain your bank account, get you fired, or even have you arrested.

Do I have your attention? Think of what would happen if someone were to get access to your PC at home or at work. They could wreak havoc on your life.

A malevolent prankster could send an email from your work email address under your name. It could be a rude note to the head of your company. It could give phony information about a supposedly imminent financial collapse of your company to the local newspaper. It could be a threat to the U.S. president, generating a visit from the Secret Service.

Here's the deal: If you have done anything with your smartphone as described in this book past Chapter 3, I expect you'll want to take steps to protect your smartphone. This is the burden of having a well-connected device. Fortunately, most of the steps are simple and straightforward.

# Using a Good Case

The Samsung Galaxy S22 is sleek and beautiful. The Galaxy S22 has a really cool design that draws attention from people walking by. Plus, the front is made of Gorilla Glass from Corning. This stuff is super-durable and scratch-resistant.

So why am I telling you to cover this all up? It's like buying a fancy dress for a prom or wedding and wearing a coat all night. Yup. It's necessary for safe mobile computing.

**WARNING**

Speaking from personal experience, dropping a Galaxy phone on concrete can break the glass and scramble some of the innards. This can happen if you simply keep your phone in a pocket.

There are lots of choices for cases. The most popular are made of silicone, plastic, or leather. There are different styles that meet your needs from many manufacturers.

You can find several good options (see Figure 17-1).

**FIGURE 17-1:**
Cases for the Galaxy S22E, S22, and S22+.

*Left: Pavel Kubarkov / Shutterstock.com; Center: doomu / Shutterstock.com; Right: Marko Poplasen / Shutterstock.com*

Otterbox is a brand that has been making cases for multiple levels of protection. The Defender Series for the Galaxy S22, its highest level of protection, is on the right in Figure 17-2; the Strada Series is on the left, and the Symmetry Series is in the center.

Strada Series     Symmetry Series     Defender Series

*Photograph courtesy of Otter Products, LLC*

**FIGURE 17-2:** Otterbox cases for the Samsung Galaxy S22, S22+, and S22 Ultra.

**TIP**

I am told by the cooler members of my clan that wearing the belt clip is the modern equivalent of wearing a pocket protector.

You don't just use a good case so that you can hand off a clean used phone to the next lucky owner. A case makes it a little less likely that you will lose your phone. Your Galaxy S22 in its naked form is shiny glass and metal, which are slippery. Cases tend to have a higher coefficient of friction and prevent your phone from slipping out of your pocket when you take a ride in an Uber car.

More significantly, a case protects your phone against damage. If your phone is damaged, you have to mail it or bring it to a repair shop. The problem is that many people who bring their phones in for repair don't wipe the personal information off their devices. You really hope that the repair shop can pop off the broken piece, pop on a new one, and send you on your way. It's rarely that easy. Typically, you need to leave your phone in the hands of strangers for some period of time. For the duration of the repair, said strangers have access to the information on your phone.

The good news is that most workers who repair phones are professional and will probably ignore any information from the phone before they start fixing it.

However, are you sure that you want to trust the professionalism of a stranger? Also, do you really want the hassle of getting a new phone? Probably not, so invest in a good case and screen cover. There are many options for different manufacturers of cases. Be sure to shop around to come up with the ideal combination of protection and style right for you.

# Putting It on Lockdown

Let's start with the basics of the Lock Screen. When you leave your phone alone for a while, or when you're doing something and you want to do something else, you can briefly press the power button to turn off the screen. This saves some power. Depending on what you want to do, you can make it easy or hard to return to using your phone by using the Lock Screen. The thinking is that you rarely lose your phone, or have it stolen when you're actively using it. It typically disappears when you leave it alone for a while.

To protect the information you keep on your phone, your S22 has the option of a Lock Screen on which, in most circumstances, you need to prove to the phone that you're you, and not some evildoer or a nosy family member. Figure 17-3 shows you a typical, and stylish, Lock Screen.

TIP

For reasons that sort of make sense, your phone uses some terminology that can be confusing. To clarify, the term *Screen Lock* is an option you can select to prevent unauthorized users from getting into your phone. The term *Lock Screen* is what you see in Figure 17-3 after enabling the Screen Lock option.

The most basic effort you can take to protect your phone is to put some kind of a screen lock on your phone. If you're connected to a corporate network, the company may have a policy that specifies what you must do to access your corporate network. Otherwise, you have seven choices, listed here in increasing degrees of security:

>> None (as in your phone will open up to whatever was there before the screen image went dark).

>> Unlock the Lock Screen with a simple swipe across the screen.

>> Unlock with a pattern that you swipe on the screen.

>> Unlock with a PIN.

>> Unlock with facial recognition.

>> Unlock with a password.

>> Unlock with your fingerprint.

Lock Icon

FIGURE 17-3:
The Lock
Screen.

You can select any of these options in the Lock Screen option in Settings. Here's what you do:

**1. Tap the Settings icon.**

This should be old hat by now.

**2. Tap the Lock Screen.**

This brings up the options seen in Figure 17-4. You set some of the options mentioned in the preceding list and others when you follow the next instruction. This can be a little confusing, but bear with me while I explain your options and tell you where to go.

**3. Tap the Screen Lock Type link.**

This brings up the options seen in Figure 17-5. Each option prompts you through what it needs before establishing your security selection.

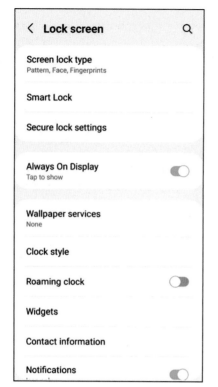

**FIGURE 17-4:**
The Lock
Screen options.

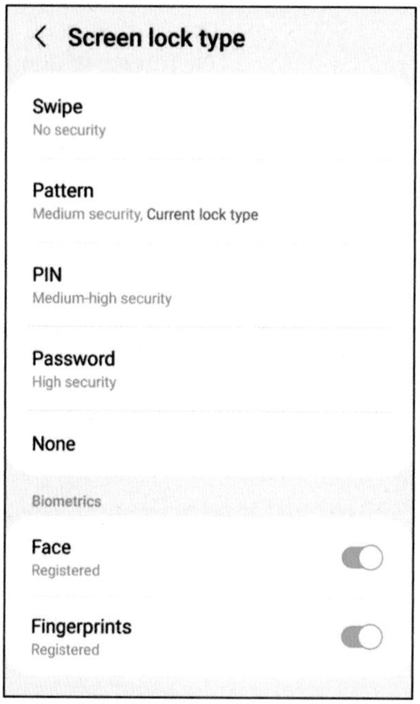

**FIGURE 17-5:**
The Screen
Lock options.

During the initial setup of your phone, you may have been asked to enter a pattern. The person at the store may have walked you through this part. If so, I hope you remember that pattern because you may need it to get to the Screen Lock options.

If you're worried about what will happen if you forget your pattern, pay very, very close attention to the section "Rescuing Your Phone When It Gets Lost," later in this chapter. In addition to helping you find a lost phone, this section covers how to set up your Samsung account from your PC — these steps will not only find your phone but also unlock it. Don't lose the password to your Samsung account — if you do, your beautiful and costly Galaxy S22 will only be useful as a paperweight.

## Preparing for your Screen Lock option

Regardless of what Screen Lock you choose, I recommend that you have ready the following choices at hand:

>> An unlock pattern

>> A PIN

>> A password

>> Your fingerprint

>> Your face

To clarify definitions, a *PIN* is a series of numbers. In this case, the PIN is four digits. A *password* is a series of numbers, upper- and lowercase letters, and sometimes special characters, and is typically longer than four characters. A PIN is pretty secure, but a password is usually more secure. Have them both ready, but decide which one you would prefer to use.

## Selecting among the Screen Lock options

The first option of skipping the whole lock screen thing altogether is your choice. Good luck with that one *<sarcasm alert!>*.

The second option, unlocking your phone with a swipe, fools exactly no one and doesn't slow anyone down. Rather than just having the Home screen appear, your phone tells you to swipe your finger on the screen to get to the Home screen. This is about as secure as waving at intruders and tossing them your phone, wallet, and keys. The only reason to choose this over the first option is that you like to see, on occasion, the abstract image shown in Figure 17-3 without the lock icon. Let's keep going.

I recommend drawing out a pattern as the minimum screen-lock option. This is quick and easy. Tap the Pattern option on the screen seen in Figure 17-6 to get started. The phone asks you to enter your pattern and then asks you to enter it again. It then asks you to enter a PIN in case you forget your pattern.

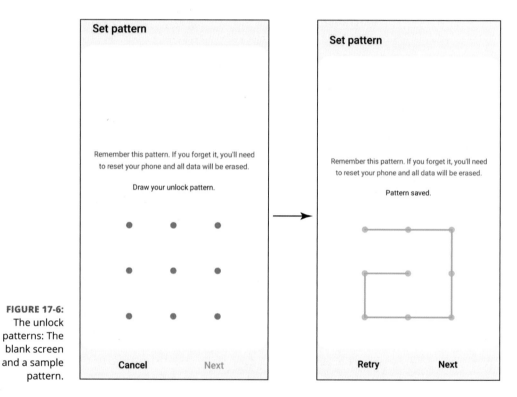

FIGURE 17-6:
The unlock patterns: The blank screen and a sample pattern.

The unlock pattern is a design that you draw with your finger on a nine-dot screen, as shown in Figure 17-6.

The image on the right in Figure 17-6 happens to include all nine dots. You do not need to use all the dots. The minimum number of dots you must touch is four. The upper limit is nine because you can touch each dot only once. As long as you can remember your pattern, feel free to be creative.

**TIP**

Be sure to use a PIN you can remember a long time from now. You need this PIN only if you forget your pattern. That is a very rare situation for most people.

Each time you enter a Screen Lock option, two things will happen:

>> If you've already entered a Screen Lock option, it may ask you to enter it. For example, if you've put in a pattern and you want to add a PIN, it may ask you to enter that pattern. This helps prevent someone from messing with your Screen Lock.

>> After you enter your PIN, password, or whatever, your phone will save the pattern to your Samsung Account. This happens automatically, which is convenient.

The next two options on the Screen Lock screen, PIN and Password, are more secure, but only as long as you avoid the obvious choices. If you insist upon using the PIN "0000" or "1111" or the word "password" as your password, don't waste your time. It's standard operating procedure within the typical den of thieves to try these sequences first. That's because so many people use these obvious choices.

**WARNING**

If, someday, you forget your pattern, your PIN, or your password, the only option is to do a complete reset of your phone back to original factory settings. In such a case, all your texts and stored files will be lost. Try to avoid this fate: Remember your pattern, PIN, or password.

## Entering your face

Facial recognition is even easier! From the Settings page, tap the Biometrics and Security link. This brings up the screen shown in Figure 17-7.

Tap the Face Recognition link. You'll be prompted to hold the phone toward your face. The recognition screens are shown in Figure 17-8. They work remarkably fast.

The last screen asks you some questions about whether you want faster facial recognition, whether you want to require that your eyes be open, and whether you want the screen to brighten in case it's dark. Sure. Go ahead and tap Done. Your phone now has your face.

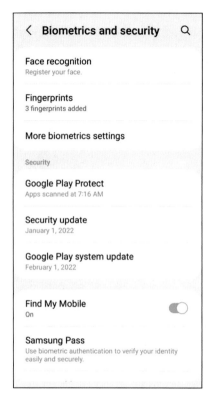

**FIGURE 17-7:**
The Biometrics
and Security
screen.

**TIP**

The facial recognition works fast and is very convenient. It's so fast and convenient at detecting your face and unlocking your phone, in fact, that you may wonder if you accidentally turned off the lock screen and made your phone totally vulnerable. You'll probably check a few times before coming to the realization that your phone is super-fast and super-smart. If it still bothers you, remove the image of your face that's stored on your phone.

## Entering your fingerprints

Using your fingerprint to unlock your phone is very convenient and very effective. There are a few hoops you need to make this happen. You need to give your phone enough views of the finger you will use, which can be whichever finger you like, so that it is sure that it is you. It also wants to make sure that you, and only you, have stored a fingerprint. Your phone can store multiple fingers. Any of your fingers will work, so have at it.

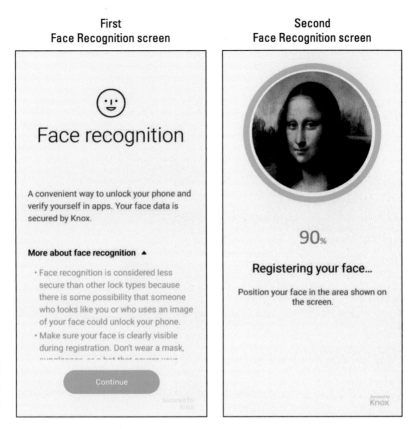

First
Face Recognition screen

Second
Face Recognition screen

😊

Face recognition

A convenient way to unlock your phone and
verify yourself in apps. Your face data is
secured by Knox.

**More about face recognition** ▲

· Face recognition is considered less
  secure than other lock types because
  there is some possibility that someone
  who looks like you or who uses an image
  of your face could unlock your phone.
· Make sure your face is clearly visible
  during registration. Don't wear a mask,

Continue

Secured by
Knox

90%

Registering your face...

Position your face in the area shown on
the screen.

Secured by
Knox

**FIGURE 17-8:**
The facial
recognition
screen for Lisa
del Gioconda.

Here's what you do:

1. **From the Settings screen, tap the Lock screen link.**

2. **Tap the Screen Lock Type link.**

    You're asked to enter your security pattern.

3. **Enter your security pattern.**

4. **Tap the Fingerprints link.**

    The leftmost screen in Figure 17-9 appears.

5. **Press your finger on the fingerprint icon on the screen.**

    Keep following the directions on the screen for how it wants you to move until
    you reach 100 percent. The screen in the middle of Figure 17-9 will eventually
    get tired of telling you to press hard, cover the sensor, and move the position
    of your finger. Before you know it, you're at 100 percent.

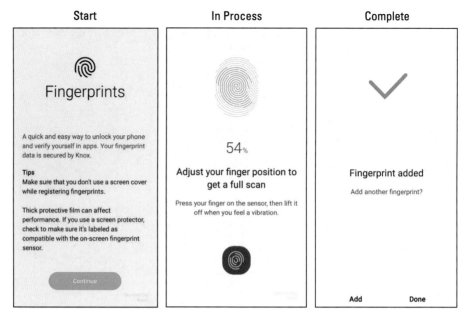

| Start | In Process | Complete |
|---|---|---|

**Fingerprints**

A quick and easy way to unlock your phone and verify yourself in apps. Your fingerprint data is secured by Knox.

**Tips**
Make sure that you don't use a screen cover while registering fingerprints.

Thick protective film can affect performance. If you use a screen protector, check to make sure it's labeled as compatible with the on-screen fingerprint sensor.

Continue

54%

**Adjust your finger position to get a full scan**

Press your finger on the sensor, then lift it off when you feel a vibration.

**Fingerprint added**

Add another fingerprint?

Add          Done

**FIGURE 17-9:** The Fingerprints registration screen.

TIP

If you ever want to add more of your fingers (up to nine more for most of us, although no judgments here), you can do so any time you want by tapping the Biometrics and Security link within Settings.

Your fingerprints are now in memory and ready to let you get to your Home page with a quick press or swipe. Give it a try. It is very slick!

# Creating a Secure Folder

Secure Folder is an option to protect data and/or applications on your device. It's an exceptionally secure option: It scrambles every file stored in the Secure Folder into gibberish, which it rapidly descrambles when you need the information. When you lock the folder, it's inaccessible and for all practical purposes invisible. When you enter your security option, like a password, the files are there for you to use like normal. This sounds great, but in practice, there are some important issues to think about.

First, if you put files in the Secure Folder and then forget your password, you can only get it back using your Samsung account. Do try not to lose both or you will never, ever, ever, ever get it back together.

Next, you already have one layer of security to get past your lock screen. If you get lazy and don't bother to lock the folder or use the same PIN or password, it's a waste of time.

If you're sure that using the secure folder is for you, here are the steps:

1. **From the Applications screen, tap the Secure Folder link.**

   Doing so brings up some welcome screens until your get to the Secure Folder home screen, shown in Figure 17-10.

**FIGURE 17-10:**
The Secure
Folder screen
and pop-up.

2. **Tap the three dots to the right of the plus sign.**

   The pop-up appears.

   The first thing you need to do is figure out what files you want to have protected. Let's say it's a picture of your baby on a bearskin rug. You want to keep it, but you don't want just anyone to see it.

3. **Tap the Add files links.**

   At this point, you see the options presented in Figure 17-11.

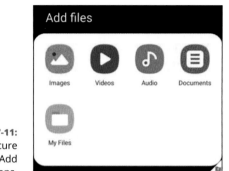

**FIGURE 17-11:**
The Secure
Folder Add
Files options.

4. **Select Images.**

   You're taken to the Gallery app.

5. **Select the image in question and tap Done.**

   You're asked if you want to copy it or move it. In most cases, you want to move it. If you choose to copy it, a version will remain in the Gallery.

6. **Select Move.**

   The image is now in your Secure Folder and gone from the Gallery.

7. **When you're ready, tap the three dots on the Secure Folder home page, bring up the pop-up, and tap the Lock and Exit link.**

   If you don't lock the Secure Folder, it's just as accessible as it was before.

When you want to access the image to chuckle at the excessive cuteness of the baby picture, you'll be prompted to use the security option in Settings. It could be a pattern, a PIN, or a password. Figure 17-12 shows the options. Just tap the lock type and then select the security option.

**REMEMBER**

You and you alone get to decide how carefully you protect the data. Choose wisely.

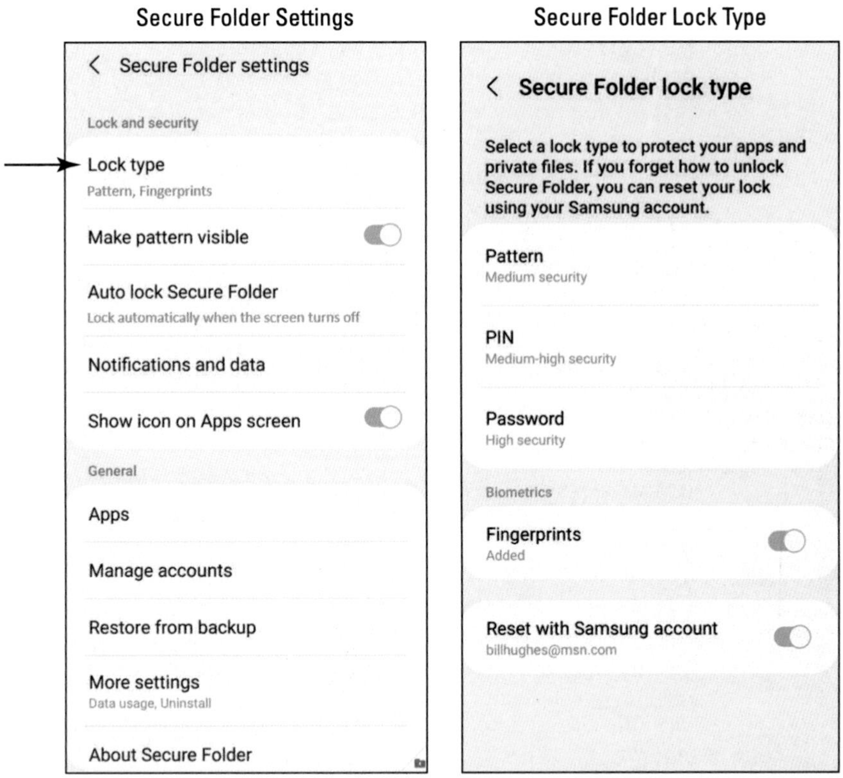

Secure Folder Settings

Secure Folder Lock Type

< Secure Folder settings

Lock and security

→ Lock type
Pattern, Fingerprints

Make pattern visible

Auto lock Secure Folder
Lock automatically when the screen turns off

Notifications and data

Show icon on Apps screen

General

Apps

Manage accounts

Restore from backup

More settings
Data usage, Uninstall

About Secure Folder

< Secure Folder lock type

Select a lock type to protect your apps and
private files. If you forget how to unlock
Secure Folder, you can reset your lock
using your Samsung account.

Pattern
Medium security

PIN
Medium-high security

Password
High security

Biometrics

Fingerprints
Added

Reset with Samsung account
billhughes@msn.com

**FIGURE 17-12:**
The Secure
Folder Settings.

# Using Knox to Make Your Phone as Secure as Fort Knox

It doesn't matter whether you bought the phone at a retail store or your company supplied it to you. The fact of the matter is that company data that resides on your phone belongs to your employer. You probably signed a document (now sitting in your HR file) that states that you agree with this arrangement.

This policy is necessary for the company because it has a financial and legal obligation to protect company data, particularly if it pertains to individual customers. Like it or not, this obligation trumps your sense of privacy over the phone you bought and (still) pay for. If this really gets under your skin, you can always carry two smartphones: one for business and one for personal use. This solves the problem, but it's a hassle.

However, there is a better way. Samsung has a highly skilled group that has developed a secure system called *Knox*. Knox logically divides your phone into two modes: one for business use and the other for your personal use. You tap an icon, and you're in business mode. You tap another icon, and you're in personal mode. Switching between the two is instant, and Knox keeps the information from each mode separate.

This is a capability that is only of interest to you if your employer offers support for the service.

When it is available, Knox comes with three capabilities for the employer:

>> Security for the Android OS

>> Limitations on the apps that access the business side of your phone

>> Remote mobile-device management

This arrangement means that your employer can remotely control the business apps and potentially wipe the data on the business side at its discretion — but it has nothing to do with your personal information. The personal side remains your responsibility.

You may want to suggest to your company's IT department that it look into supporting Knox. Doing so can take a burden off your back.

# Being Careful with Bluetooth

In Chapter 3, I cover syncing your phone with Bluetooth devices. I don't mention the potential for security risk at that point. I do it now.

Some people are concerned that people with a radio scanner can listen in on their voice calls. This was possible, but not easy, in the early days of mobile phone use. Your Galaxy S22 can use only digital systems, so picking your conversation out of the air is practically impossible.

Some people are concerned that a radio scanner and a computer can pick up your data connection. It's not that simple. Maybe the NSA could get some of your data that way using complicated supercomputing algorithms, but it's much easier for thieves and pranksters to use wired communications to access the accounts of the folks who use "0000" as their PIN and "password" or "password1" as their password.

Perhaps the greatest vulnerability your phone faces is called *bluejacking*, which involves using some simple tricks to gain access to your phone via Bluetooth.

Do a test: The next time you're in a public place, such as a coffee shop, a restaurant, or a train station, turn on Bluetooth. Tap the button that makes you visible to all Bluetooth devices and then tap Scan. While your Bluetooth device is visible, you'll see all the other Bluetooth devices in your vicinity. You'll probably find lots of them. If not, try this at an airport. Wow!

If you were trying to pair with another Bluetooth device, you'd be prompted to see whether you're willing to accept connection to that device. In this case, you are not.

However, a hacker will see that you are open for pairing and take this opportunity to use the PIN 0000 to make a connection. When you're actively pairing, your Bluetooth device won't accept an unknown device's offer to pair. But if your device is both unpaired and visible, hackers can fool your Bluetooth device and force a connection.

After a connection is established, all your information is available to the hackers to use as they will. Here are the steps to protect yourself:

>> **Don't pair your phone to another Bluetooth device in a public place.**
Believe it or not, crooks go to public places to look for phones in pairing mode. When they pair with a phone, they look for interesting data to steal. It would be nice if these people had more productive hobbies, like Parkour or searching for Bigfoot. However, as long as these folks are out there, it is safer to pair your Bluetooth device in a not-so-public place.

>> **Make sure that you know the name of the device with which you want to pair.** You should pair only with that device. Decline if you are not sure or if other Bluetooth devices offer to connect.

>> **Shorten the default time-out setting.** The default is that you will be visible for two minutes. However, you can go into the menu settings and change the option for Visible Time-out to whatever you want. Make this time shorter than two minutes. *Don't set it to Never Time Out.* This is like leaving the windows open and the keys in the ignition on your Cadillac Escalade. A shorter time of visibility means that you have to be vigilant for less time.

>> **From time to time, check the names of the devices that are paired to your device.** If you don't recognize the name of a device, click the Settings icon to the right of the unfamiliar name and unpair it. Some damage may have been done by the intruder, but with any luck, you've nipped it in the bud.

Here's an important point: When handled properly, Bluetooth is as secure as can be. However, a few mistakes can open you up to human vermin with more technical knowledge than decency. Don't make those mistakes, and you can safely enjoy this capability, knowing that all the data on your phone is safe.

# Protecting against Malware

One of the main reasons application developers write apps for Android is that Google doesn't have an onerous preapproval process for a new app to be placed in the Play Store. This is unlike the Apple App Store or Microsoft Windows Phone Store, where each derivation of an app must be validated.

Many developers prefer to avoid bureaucracy. At least in theory, this attracts more developers to do more stuff for Android phones.

However, this approach does expose users like you and me to the potential for malware that can, inadvertently or intentionally, do things that are not advertised. Some of these "things" may be minor annoyances, or they could really mess up your phone (for openers).

Market forces, in the form of negative feedback, are present to kill apps that are badly written or are meant to steal your private data. However, this informal safeguard works only after some poor soul has experienced problems — such as theft of personal information — and reported it.

Rather than simply avoiding new apps, you can download apps designed to protect the information on your phone. These are available from many of the firms that make antivirus software for your PC. Importantly, many of these antivirus applications are free. If you want a nicer interface and some enhanced features, you can pay a few dollars, but it isn't necessary.

Examples include NQ Mobile Security and Antivirus, Lookout Security and Antivirus, Kaspersky Mobile Security, and Norton Security Antivirus. If you have inadvertently downloaded an app that includes malicious software, these apps will stop that app.

# Downloading Apps Only from Reputable Sources

Another way to avoid malware is to download mobile software only from trustworthy websites. This book has focused exclusively on the Google Play Store. You can download Android apps for your phone from a number of other reputable sites, including Amazon Appstore and GetJar.

Keep in mind that these stores are always on the lookout to withdraw applications that include malicious software. Google uses an internally developed solution it calls Bouncer to check for malicious software and remove it from the Play Store. Other mobile software distribution companies have their own approaches to addressing this problem. The problem is that policing malicious software is a hit-or-miss proposition.

As a rule, you should hesitate to download an Android app unless you know where it has been. You are safest if you restrict your app shopping to reputable companies. Be very skeptical of any other source of an Android app.

# Rescuing Your Phone When It Gets Lost

Other options allow you to be more proactive than waiting for a Good Samaritan to reach out to your home phone or email if you lose your phone.

There are apps that help you find your phone. Here are a few "lost it" scenarios and some possible solutions for your quandary:

>> **You know that you lost your phone somewhere in your house. You would try calling your own number, but you had your phone set to Vibrate Only mode.**

  *Solution:* Remote Ring. By sending a text to your phone with the "right" code that you preprogrammed when you set up this service, your phone will ring on its loudest setting, even if you have the ringer set to Vibrate Only.

TIP

  If you know that your phone is in your house, the accuracy of GPS isn't savvy enough to tell you whether it's lost between the seat cushions of your couch or in the pocket of your raincoat. That's where the Remote Ring feature comes in handy.

>> **You lost your phone while traveling and have no idea whether you left it in a taxi or at airport security.**

*Solution:* Map Current Location. This feature allows you to track, within the accuracy of the GPS signal, the location of your phone. You need access to the website of the company with which you arranged to provide this service, and it will show you (on a map) the rough location of your phone.

Here is where having an account with Samsung comes in handy. Hopefully, you signed up for a Samsung Account when you first got your phone. If you did, you're signed up for the Find My Mobile at `http://findmymobile.samsung.com`. Figure 17-13 shows the Find My Mobile PC screen.

**FIGURE 17-13:** The Samsung Find My Mobile PC screen.

You can see from this screen that my phone is at 935 Pennsylvania Ave. NW, Washington, DC, which happens to be the headquarters for the FBI. All you need to do is get to a PC and sign in to your Samsung account (unless your phone is at the FBI). You can tell your phone to ring by clicking on Ring My Device. You can have the PC bring up a map by clicking Locate My Device.

I suggest trying these out before you lose your phone the first time.

# Wiping Your Device Clean

As a last-ditch option, you can use Find My Mobile (see preceding section) to remotely disable your device or wipe it clean. Here are some of the possible scenarios:

>> **You were robbed, and a thief has your phone.**

*Solution:* Remote Lock. After your phone has been taken, this app allows you to create a four-digit PIN that, when sent to your phone from another mobile phone or a web page, locks down your phone. This capability is above and beyond the protection you get from your Screen Lock and prevents further access to applications, phone, and data.

**WARNING**

If you know that your phone was stolen — that is, not just lost — do *not* try to track down the thief yourself. Get the police involved and let them know that you have this service on your phone — and that you know where your phone is.

>> **You're a very important executive or international spy. You stored important plans on your phone, and you have reason to believe that the "other side" has stolen your phone to acquire your secrets.**

*Solution:* Remote Erase. Also known as Remote Wipe, this option resets the phone to its factory settings, wiping out all the information and settings on your phone.

**TIP**

You can't add Remote Erase *after* you've lost your phone. You must sign up for your Samsung service beforehand. It's not possible to remotely enable this capability to your phone. You need to have your phone in hand when you download and install either a lock app or a wipe app.

# Chapter **18**

# Ten Features to Look for Down the Road

With the power of your Samsung Galaxy S22 and the flexibility offered in Android applications development, it can be difficult to imagine that even more capabilities could be in the works. In spite of this, the following are ten features that would improve the usability and value of your Galaxy S22 phone.

## Your Medical Information Hub

All kinds of fitness apps track your steps. There are also medical apps that let you access the information for your doctor and access the results of medical tests. But these apps are just the tip of the iceberg.

Many companies make devices that connect to your smartphone over Bluetooth. Examples include blood pressure cuffs, continuous positive airway pressure (CPAP) machines, pulse oximeters, and glucose monitors. These devices collect the information and send it securely to an application on the Internet so you or medical professional can monitor trends.

The ultimate solution are medical sensors that you wear and alert you, your loved ones, or medical/emergency response professionals with your location and situation. This technology can be tricky, because a reading that is fine for you may signal a problem for someone else. The technology challenges include collecting data from the different kinds of sensors and having the intelligence, probably artificial, to know whether this condition is serious or urgent and whom to call.

Your phone can also be a convenient way for you to present information on your vaccination status and carry other useful information on your medical status. The simplest option is to take a picture of the information and store it in your photo gallery. This may be sufficient for some uses, such as showing the staff at a restaurant that you meet the vaccination standards set by the local board of health. But using the photo gallery to store your sensitive personal health information may not have the level of security you want or accessibility that you need if you aren't able to give emergency crews your health data. Some apps try to solve this issue, but they haven't gained widespread acceptance.

# Better 911 Services

The 911 system has been keeping the United States safe for more than 45 years. But the dirty secret of this service is that the underlying technology used to communicate information on the caller's location hasn't been updated in a long time.

To put this in perspective, your smartphone is designed to work with data at up to 300 million bits per second. When you call 911, the phone that answers your call is designed to work with data at up to 120 bits per second. (Seriously. I am not making this up.)

Many states and regions are trying to address this problem. This effort is called next-generation 911, or NG911. NG911 promises to make the information you already have on your phone available to the people who are sending you help. This new technology is slowly being implemented region by region in different states and counties.

With a larger data pipeline between your smartphone and the first responders, you can send anything that's relevant, including medical history, your emergency contacts, insurance data, and whether you have any protection orders against stalkers. All this information can help you — and it's available right away because, regardless of where you happen to be, your phone is typically there with you.

The larger pipeline also allows you to send information. Let's say you're visiting a new city and you see the drapes on fire in a multistory building. You call 911, but you don't know where you are or the address of the building, and all you can do is guess the floor with the fire. This is next to worthless for the 911 operator. With NG911, you can point the camera of your phone to show the operator the fire. Now the operator knows where you are and can pinpoint the fire based on the orientation of the phone. Problem solved and lives saved.

# Home Internet of Things Services to Differentiate Real Estate

Much of the discussion about home Internet of Things (IoT) services involves individuals buying new appliances, thermostats, lighting, security, and other items. That works for homeowners. It can also work for apartment dwellers and new home buyers where the landlord or builder provides these services as a way to differentiate their offerings. Typically, they offer a mobile app to access all this stuff. This makes it easier for you to adopt this capability without having to maintain it yourself.

**TECHNICAL STUFF**

The *Internet of Things* is a catchall term for when a manufacturer gives you, the customer, the ability to control a "thing" from your phone, like use your phone as a TV remote control. Unlike your TV or stereo remote that uses an infrared light beam to control your TV or stereo, the manufacturer offers you the ability to control your thermostat, lights, or doorbell through the Internet using an app on your S22. Unlike the TV remote, it doesn't matter if you're in the room. You can check in on the electronic "thing" from anywhere you happen to be.

This also works at the office (if you still go into the office, that is). If your office space is too hot or too cold, you can submit a ticket and maybe the building staff will adjust the thermostat in a few hours. In the future, your office will give you an app on your smartphone that will replace your security card, let you tell them when your office is too hot or too cold with the tap of an icon, remind you to keep safe distances, and let you reserve conference rooms.

# New Delivery Concepts

Most of us over the age of 5 years old know our address. It is on our driver's license. It is where we get things delivered from Amazon.com and other e-tailers. It is customary to have stuff shipped to a real address. However, if you are homeless, a road-warrior for business, or simply social, you are not always at that physical address.

Here is the future: Your smartphone will increasingly be used as the place to arrange for delivery. You can arrange to have physical stuff delivered to where you are, rather than an address. Shippers know where you will be based upon your smartphone, and can deliver to you personally.

This delivery can be by a drone, but more likely, it will still be by truck. They can find you and get the physical whatever-it-is-that-you-want in your hands quicker than you can say Jack Robinson (in other words, impressively fast).

# Smarter Customer Care for Your Phone

You may not realize this, but your cellular carrier lives on pins and needles during the first few weeks after you get your new phone. This is the period when you can return your phone, no questions asked. Once you go past that date, you cannot cancel your contract without a lot of hassle on your part.

This is why, if you bring your phone back to the store reporting a problem, your carrier will tend to swap out your old phone for a brand new one.

Usually, you'll just take the new phone and walk out with a big smile on your face. This outcome is good customer care for most products — but not necessarily good customer care for smartphones. The reason? One of the most common sources of trouble has nothing to do with the phone at all. For example, you may have left your Wi-Fi and Bluetooth on all the time, causing your battery to drain too fast. You may have left your phone on your dashboard and cooked the battery. Or, through no fault of your own, you may have downloaded two badly written apps that conflict with each other, causing the CPU in the phone to run nonstop as these two apps battle it out for resources. The problem here is that unless someone spends the time to help you with the underlying problem, you'll be back in the store with the same problem.

At that point, however, you cannot return your phone. If you're sympathetic, or very annoying, your carrier may give you a refurbished phone. You walk out of the store, but without the biggest smile on your face. Unfortunately, nobody dealt with the underlying trouble, so you'll be back, once again, with the same problem.

No surprise if you start to think that the problem is with that darn phone. In fact, the store needs to listen to you about how you're using the phone and then help you get the most out of it. This is hard to do in a retail environment where the sales force is under pressure to sell lines of service and gets no concrete reward for helping you with your problem.

This is where smarter customer care comes in. With the proper tools, you can work with a product expert to troubleshoot your phone. Some companies specialize in understanding the underlying problems and coming up with solutions for consumers. This kind of customer care costs the carriers a little more, but it makes for fewer unnecessary returns of perfectly good phones — and for much happier customers.

# Smartphone as Entertainment Hub

Today's savvy technology customer has some or all of the following:

>> Intelligent Big Screen TV or Dumb TV with a set-top box for home entertainment.

>> Home Stereo System for immersive music.

>> Home PC/Laptop for home management and entertainment.

>> Work or School PC/Laptop. This helps you keep home and work separate.

>> Tablet for the folks who find the larger screen to complement their smartphone.

>> Gaming Console for games.

It would be hard to argue that having all these devices in one would not have advantages except for the following realities:

>> Docking your smartphone to your big-screen TV and/or stereo system is not very convenient.

>> Docking your smartphone so that you can use a full-size computer screen, keyboard, and mouse is also not very convenient.

>> The tablet screen size can be easier to view.

>> The digital music on your smartphone is great, but there are many people who still want access to their DVD, CD, and/or vinyl collections.

>> Gaming Consoles still have better games than what are available on a smartphone.

There will come a day when docking becomes so convenient that all your computing and entertainment will come through your smartphone, and you'll connect to an ergonomic keyboard, a bigger screen, and/or more powerful music amplifier at your convenience. The elements are there today, but they are just not convenient enough.

# Driving in Your Car

After years of tempting us in science fiction movies, driverless cars are now starting to appear on the roads. Several insurance companies offer you an app to store your auto insurance information. Some car manufacturers allow you to check the location of your car and lock/unlock your car. These are all a good start.

What about the car unlocking and adjusting the seats and mirrors when you get close without one of those bulky key fobs? What about communication between your car and phone that would more intelligently let you use certain phone features if you are in the passenger seat rather than the driver seat or, if you are in the driver seat, to give you the freedom to check some texts while the traffic is stopped?

It is generally preferable for a car to have its Intelligence built in and run off the car battery. That said, your smartphone knows more about you and your preferences. The best of both worlds is when your car and your phone can collaborate to provide the best and safest experience.

# Serving You Better

The smartphone is the mechanism that companies use to find a better way to serve you. This will show up in a few ways.

The first is in much better mobile advertising. There has been talk for a long time that advertisers can tell where you are and then give you ads or coupons based upon your proximity. This is just now starting to become a reality. It's still kinda clumsy. If you have been Googling, say, a new matteress, you may see ads relating to sales about bedding options. In the future, however, your phone may tell you about yet another sales event going on for mattresses when you walk by any of the high-end bed stores in the mall. That is cool.

Even cooler is to be told that that particular model is on sale at a mattress store near a major mall and that this particular grand opening and going-out-of-business sale is the best in town by over $200. This can happen if you are doing research at the moment. It may be possible for your app to flag you if you have ever done this kind of research on the Internet.

The second way smartphones are enhancing service is by automating the order-taking process. For example, wherever you see a video kiosk, that company can allow you to interact with the information on the kiosk with your Samsung Galaxy S22. Some fast-food restaurants have installed kiosks so that customers can bypass the line to order from the menu. The items are totaled and paid for by credit card. The customer steps aside and waits for their number to be called.

This exact transaction could take place on your smartphone. You don't have to wait while the guy in front of you struggles to decide between a double cheeseburger or two single cheeseburgers . . . and hold the pickles. Instead, you can place your order on your smartphone and wait with eager anticipation to find out what toys are in your kid's meal.

# Placing You Indoors

The Global Positioning System (GPS) is great. It can tell you precisely your location on the road and off the road. Chapter 14 covers how it can help you get to where you want to be.

There is a confession to be made. GPS doesn't work so well within buildings. Once you are in a building, your phone only kind of knows where you are. The satellite responsible for sending out signals so that your phone can figure out your position is no longer visible, and the GPS in your phone waits patiently until it sees another satellite. Until then, it stops trying to figure out where you are.

This is rarely a problem if you are, say, at your home. It is probably only a few thousand square feet. Things are altogether different if you are wandering around the Palazzo Casino in Las Vegas. At 6.95 million square feet, this building has the most floor area of any building in the United States.

You could be anywhere in the 160 acres of luxury resort space. It is unfortunate if the Carnevino Steakhouse sends you a mobile coupon offering a two-for-one meal to your Galaxy S22 while you are already sitting in the Delmonico Steakhouse. It can be a tragedy if you start choking on too big a bite of your porterhouse steak and emergency services cannot find you.

There are some efforts out there to come up with a better way to locate you when you are indoors. One approach has been to estimate your location based upon signal strength of known Wi-Fi access points. Companies that hire a large number of smarticles (smart people) are trying a variety of technologies to address this problem. May the best company win so that we may all get better location-based services!

# Reducing Your Carbon Footprint

The concept of a "smart home" is gaining momentum for the consumer electronics. So, what is a smart home and what's in it for you? The basic idea is that appliances, doorbells, computers, and lighting connect to a central hub that monitors everything and allows you to control stuff from your smartphone. It sends you an alert if someone rings your doorbell and shows you who it is. You can turn lights on or off. It can turn your thermostat up or down based upon your daily routines.

The next step beyond remote control is even smarter courtesy of your S22. It will not only be able to tell when you leave the house, but also know your location and direction and how soon you'll be home. In addition, the central hub will monitor your use of electricity and be able to tell what appliance is in use. It can tell the difference between the power use for your oven, the hot water heater, and the television.

So if you forgot to turn off the iron on the ironing board as you drove away from home, no problem. The hub sends you a text. You tell the hub to turn off the iron from your Galaxy S22. When the hub sees you're gone, it adjusts the thermostat to require less power. When the hub sees that you're on your way home, your smart home readjusts the thermostat so that everything is comfortable when you walk in the door. Your smart home is no longer making its best guesses based upon what it thinks is your normal routine. Your smart home knows exactly where you are and how soon you'll be walking in the door because your trusty smartphone let the hub know. This all reduces your power consumption without any inconvenience to you.

# Index

# G

# About the Author

**Bill Hughes** is an experienced marketing strategy executive with more than three decades of experience in sales, strategic marketing, and business development roles at several leading corporations, including Guidehouse, Xerox, Microsoft, IBM, GE, Motorola, and US West Cellular.

Bill has worked with Microsoft to enhance its marketing to mobile applications developers. He also has led initiatives to develop new products and solutions with several high-tech organizations, including Sprint, Motorola, SBC, and Tyco Electronics as a principal consultant with Phlogiston, Inc.

Bill was a professor of marketing at the Kellogg School of Management at Northwestern University, where he taught business marketing to graduate MBA students.

Bill also has written articles on wireless technology for several wireless industry trade magazines and contributed to articles in *USA Today* and *Forbes*. These articles were based upon his research reports written for In-Stat, where he was a principal analyst, covering the wireless industry, specializing in smartphones and business applications of wireless devices.

Bill graduated with honors with an MBA degree from the Kellogg School of Management at Northwestern University and earned a bachelor of science degree with distinction from the College of Engineering at Cornell University, where he was elected to the Tau Beta Pi Engineering Honorary.

# Dedication

I would like to dedicate this book to Chris Jones and Jim DeBelina. Thank you for your mentorship.

# Author's Acknowledgments

I need to thank a number of people who helped make this book a reality. First, I would like thank my literary agent, Carole Jelen, of Waterside Publishing, for her support, encouragement, knowledge, and negotiation skills.

I would also like to thank the team at John Wiley & Sons, Inc.: Kelsey Baird, Kristie Pyles, and Elizabeth Kuball. Your expertise helped me through the creative process. Thanks for your guidance.

I would also like to thank Kristen Tatti at Otterbox, Ami Brannon at Neuvana, Cassie Pineda at Belkin, Maria Tullgren at Audio Pro, Nicole Martin at Skinit, Joseph Kawar at SlipGrips, and multiple individuals at Edelman.

Thanks go out to numerous people at Samsung who helped make this book possible.

Finally, I would like to acknowledge Ellis, Arlen, and Quinlan for enabling me to be an insufferably proud father.

## Publisher's Acknowledgments

**Associate Acquisitions Editor:** Kelsey Baird

**Project Editor:** Elizabeth Kuball

**Copy Editor:** Elizabeth Kuball

**Proofreader:** Debbye Butler

**Production Editor:** Tamilmani Varadharaj

**Cover Image:** Courtesy of William M. Hughes; Background image: © Gokcemim/Getty Images